TECHNOLOGY, INNOVATION and POLICY 12

Series of the Fraunhofer Institute
for Systems and Innovation Research (ISI)

TECHNOLOGY, INNOVATION and POLICY

Series of the Fraunhofer Institute
for Systems and Innovation Research (ISI)

Knut Koschatzky · Marianne Kulicke
Andrea Zenker (Editors)

Innovation Networks

Concepts and Challenges
in the European Perspective

With 25 Figures and 28 Tables

Physica-Verlag

A Springer-Verlag Company

Dr. Knut Koschatzky
Dr. Marianne Kulicke
Andrea Zenker
Fraunhofer Institute
for Systems and Innovation Research
Breslauer Straße 48
76139 Karlsruhe
Germany

E-mails:
ko@isi.fhg.de
mt@isi.fhg.de
az@isi.fhg.de

ISSN 1431-9667
ISBN 3-7908-1382-6 Physica-Verlag Heidelberg New York

Cataloging-in-Publication Data applied for
Die Deutsche Bibliothek – CIP-Einheitsaufnahme
Innovation networks: concepts and challenges in the European perspective; with 28 tables / Knut
Koschatzky ... (ed.). – Heidelberg; New York: Physica-Verl., 2001
   (Technology, innovation and policy; 12)
   ISBN 3-7908-1382-6

Physica-Verlag Heidelberg New York
a member of BertelsmannSpringer Science+Business Media GmbH

© Physica-Verlag Heidelberg 2001
Printed in Germany

Softcover design: Erich Kirchner, Heidelberg
SPIN 10796514   88/2202-5 4 3 2 1 0 – Printed on acid-free paper

Preface

N iA

In November 1999, the department "Innovation Services and Regional Development" of the Fraunhofer Institute for Systems and Innovation Research, Karlsruhe, carried out its second annual conference. After dealing with innovation financing for small and medium-sized enterprises in the first conference, this year's subject was innovation networking. Many theoretical and empirical contributions from different disciplines showed that co-operation between different partners is an essential prerequisite for successful innovation. The lonely inventor, who brings his invention to the market by founding his own business, the scenario Schumpeter had in mind in his early work, is far from being realistic. Only by co-operation, and thus networking, can so far unused economic and innovative potentials be exploited. While the scientific community made many theoretical and empirical contributions to the explanation of networks, many questions still remain open. For example: how can networks, if they do not emerge on their own, be initiated; how can fragmentation in innovation systems be overcome, and how can networking experiences from market economies be transferred to the emerging economies of Central and Eastern Europe? Policy-makers need to know more about the success factors of networking and network promotion, even though network stimulation is already on the political agenda.

Policy-makers' interest in these issues was documented by the fact that the Federal Ministry for Education and Research (BMBF) sponsored this conference. The participation of several representatives of different Central and Eastern European Countries was made possible by a grant from the Volkswagen Foundation. This financial support contributed significantly to the successful realisation of the conference. Thanks to the help of many people involved in organising this event, it was possible for the approximately 150 participants to spend two interesting days in Karlsruhe.

We thank the authors for their contributions, but special thanks go to Susanne Winter-Haitz and Christine Schädel with whose invaluable help we were able to compile this volume. Since most of the authors revised their paper after the conference and included discussion remarks, the volume not only offers a selection of conference papers, but gives the reader an impression of the different subjects which were presented and discussed during the two days in Karlsruhe.

Karlsruhe, November 2000

Knut Koschatzky   Marianne Kulicke   Andrea Zenker

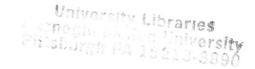

Contents

Page

4. Innovation Processes and the Role of Knowledge-Intensive Business Services (KIBS)

Simone Strambach

5. Institutions of Technological Infrastructure (ITI) and the Generation and Diffusion of Knowledge

Antoine Bureth, Jean-Alain Héraud

SECTION III: INNOVATION NETWORKS IN TRANSITION

9. Integration through Industrial Networks in the Wider Europe: An Assessment Based on Survey of Research

Slavo Radosevic

10. East German Industrial Research: Improved Competitiveness through Innovative Networks

Franz Pleschak, Frank Stummer

14.   Innovation Networks and Regional Policy in Europe

Mikel Landabaso, Christine Oughton, Kevin Morgan

List of Figures

List of Tables

Abbreviations

BAV	Business Angels Venture
BJTU	Beteiligungskapital für junge Technologieunternehmen (Business Investment Capital for New Technology-Based Firms)
BMBF	Bundesministerium für Bildung und Forschung (Federal Ministry for Education and Research)
BVK	Bundesverband deutscher Kapitalbeteiligungsgesellschaften - German Venture Capital Association e.V. (BVK)
CEE	Central and Eastern Europe
CEECs	Central and Eastern European Countries
CRITT	Centre régional d'innovation et de transfert de technologie
DFG	Deutsch Forschungsgemeinschaft (German Research Council)
DG	Directorate-General
EC	European Commission
EDP	Electronic data processing
EFRE	Europäischer Fonds für regionale Entwicklung (European Regional Development Fund)
ERDF	European Regional Development Fund
ERIS	European Regional Innovation Survey
EU	European Union
FDI	Foreign Direct Investments
FhG-ISI	Fraunhofer Institute for Systems and Innovation Research
GDP	Gross Domestic Product
GDRs	Global Depositary Receipts
GREMI	Groupe de Recherche Européen sur les Milieux Innovateurs
ICT	Information and Communication Technology
IER	Institute for Economic Research
IMA	Institut für Materialforschung und Anwendungstechnik GmbH
IR	Innovation Services and Regional Development
ISO	International Organisation for Standardisation
IT	Information Technology
ITI	Institutions of Technological Infrastructure
JVs	Joint Ventures
KIBS	Knowledge-Intensive Business Services
KISSIN	Knowledge intensive services in Europe
LFR	Less Favoured Regions
M&A	Mergers and Acquisitions
MBG	Mittelständische Beteiligungsgesellschaft (Middle Class Venture Capital Company)
MIT	Massachusetts Institute of Technology
MNCs	Multinational Companies
MNE	Multinational Enterprises
NACE	Statistical Classification of Economic Activities

NIS	National Innovation Systems
NTG	Network for Technological Expertise
NUTS	Nomenclature des unités territoriales statistiques
OECD	Organisation for Economic Cooperation and Development
OEM	Original Equipment Manufacturer
OPT	Outward Processing Traffic
R&D	Research and Development
R&TD	Research and Technological Development
R&TDI	Research, Technological Development and Innovation
RINNO	Regional prosperity through Innovation help from the European Commission
RIS	Regional Innovation Strategies
RIS	Regional Innovation System
RITTS	Regional Innovation and Technology Transfer Strategies
RKW	Rationalisierungskuratorium der deutschen Wirtschaft (Rationalisation Curatorium for German Industry)
RTD	Research and Technological Development
RTP	Regional Technology Plan
S&T	Science and Technology
SIA	Slovenian Innovation Agency
SIC	Standard Industrial Classification
SME	Small and Medium-Sized Enterprises
SOEs	State-owned enterprises
SÖSTRA	Institut für Sozialökonomische Strukturanalysen Berlin
SURS	Slovenian Office for Statistics
TGZ	Technologie- und Gründerzentren (Technology and Incubator centres)
TOU	Technologieorientierte Unternehmensgründungen (New Technology-Based Firms)
TOU-NBL	Technologieorientierte Unternehmensgründungen in den neuen Bundesländern (New Technology-Based Firms in the New Federal States)
TIPIK	Technology and Infrastructures Policy in the Knowledge-based Economy
TRIPS	Trans Regional Innovation Projects
TSER	Targeted Socio-Economic Research
UK	United Kingdom

Section I: Introduction to the Subject

1.   Networks in Innovation Research and Innovation Policy – An Introduction

Knut Koschatzky

1.1   Introduction

Networks deal with people, with firms and other institutions, and with their social and economic interaction. Economic and social networks are, of course, affected by physical networks like the Internet, telecommunication and traffic networks. Although important, the latter are not the subject of further discussion here. Social and economic networking is not a new phenomenon, but the basis of early trade which developed several thousand years ago, and, perhaps even more important, of urbanisation, the root of modern civilisation. In recent years, networking became an important subject in deductive and inductive theoretical disquisitions as well as in policy actions.

In comparison with past contributions made, for example, by Alfred Marshall at the end of last century explaining the advantages of an industrial atmosphere, fuelled by network relations between the different actors of a production system, nowadays networks are not seen as mechanistic elements in a production function, but as a contributive factor for learning and knowledge generation.

Several disciplines contributed and contribute to a better understanding of the evolution, structural characteristics, specific aspects, advantages and disadvantages of collaborative links. Sociology, political sciences, innovation and knowledge economics, evolutionary economics and regional sciences are but a few which must be mentioned. While in sociology behaviourist aspects of communication processes between different social groups are analysed, and the political sciences treat networks as an institutional platform of political influence and power (among many other aspects), major contributions to networking research were also made by evolutionary innovation economics. Since the innovation process is no longer regarded as a linear process in which some amounts of input are transformed in a kind of black box to a certain output, but as a learning process characterised by uncertainty and risk, in which several actors interact with each other, networking is believed to be a major success factor for economic development and innovation.

Nevertheless, networking does not only have positive effects. When networks are closed and the network participants are prevented from adjusting to outside needs and challenges, then these lock-in situations may cause decline, poverty and social unrest (Grabher 1993). These processes can be observed especially in states, regions or markets which are isolated or characterised by monopolistic or strong oligopolistic structures. Open networks, on the other hand, are subject to continuous change and competition, and act like a market place in which only the competitive actors survive.

The general advantage of networking in innovation processes which result from the acquisition of complementary information, knowledge and financial resources has not only made networks a subject of broad analytical treatment and research, but also an interesting option in innovation and technology policy. This applies to Germany, the European Union and many other countries as well. By stimulating contacts and co-operation between the different actors in innovation systems, synergies can be achieved and innovation potential be exploited in existing and in new firms, in research and in society as a whole. Since not all knowledge can be transferred in a codified way, but is tacit and has to be communicated by personal contacts, and co-operation in innovation is mostly built on trustful relationships between the different partners, spatial proximity plays an important role, at least in some kinds of networks. Networking and its policy support can therefore contribute not only to improved economic and technological competitiveness, but also to regional development.

Taking the still growing importance of innovation networks as a starting point, it is the objective of this volume to discuss innovation networking from different viewpoints:

- the regional viewpoint,
- the contribution of innovation networking to economic transition in Central and Eastern European Countries,
- the knowledge and learning function of innovation networks, and
- the role networking plays for innovation and firms' financing.

Before an overview on the contents of this volume is given, a short review is presented about the theoretical foundations and political implications of innovation networking.

1.2   Theoretical Aspects of Innovation Networks

The importance of interacting in innovation processes makes it clear that networking is an essential means of knowledge exchange and learning. Networks bring actors, resources and activities together and are thereby to be regarded as a system (Casti 1995: 5). They can be explained by means of transaction-cost economics and institutional economics, as well as by the different views of networking economics. According to institutional economics, networks are both a hybrid consensual form of transaction, which are settled between market and hierarchy (Williamson 1975, 1985), as well as institutional arrangements with the objective of resources acquisition and resources division (Powell 1990). A substantial difference is made between the transaction-costs and the network economics interpretation of network relations. Network economics define a *network* by a long-term relationship of different partners who co-operate on the same hierarchical level in an environment of mutual understanding and trust, while market transactions are characterised by temporary, non-lasting interactions, mostly regulated by contracts (Karlsson/Westin 1994: 3). The emergence of networks is therefore explained not primarily by cost aspects, but by strategic interests, the wish for appropriability and realising synergetic effects resulting from technological and other complementaries (Freeman 1991: 512).

Innovation networks are a specific mode of this arrangement. They are understood as all organisational forms between market and hierarchy which serve for information, knowledge and resources exchange and which help to implement innovations by mutual learning between the network partners (Fritsch *et al.* 1998). A distinction is made between vertical networks with customers and suppliers, which are usually strongly embedded in the production and value-added chain, and horizontal networks with other enterprises from the manufacturing and service sector, research establishments and other organisations (e.g. information and transfer agencies, venture capital institutions). In horizontal networks, firms have a much higher degree of freedom in selecting their partners. Networks can support both informal information exchange and the joint implementation of innovation projects, as well as pilot applications and market introduction.

The advantage of networks lies in the acquisition of complementary resources, which an individual actor does not have at his own disposal. Thus external effects can be realised by networking, which are particularly pronounced if the individual network participants are connected by horizontal, less-hierarchical and trustful relationships. According to this view innovations can only be implemented by co-operation between different (regional) actors (DeBresson/Amesse 1991). The ability to search for appropriate partners, and of utilising external, innovation-relevant knowledge depends on the absorptive capacity of a firm (Cohen/Levinthal 1990: 128). The higher the absorptive capacity, the more firms are able to seek out co-operation partners and to co-operate within network relations, not only within their regional environment, but on an international scale.

Ideally, networks are characterised by the following features (DeBresson/Amesse 1991; Fritsch 1992; Granovetter 1982; Semlinger 1998):

- trust between the participants,
- relations usually designed in a long-term time perspective,
- redundancies within the network, i.e. options and absence of hierarchy,
- openness, dynamics and flexibility,
- competition between the network actors,
- independence and voluntary co-operation,
- scale economics through co-operation.

During recent years, networking research has dealt particularly with the relevance of systematic learning in the innovation process (Coombs/Hull 1998; Cowan/Foray 1997; Cimoli/Dosi 1996; Nonaka/Takeuchi 1995; Spender 1996). It is argued that knowledge can be acquired only by systematic learning and forgetting. Learning within enterprises can be implemented at different levels (Reid/Garnsey 1998), whereas learning by interacting between customers and producers, competitors, and other enterprises as well as research establishments substantially affects innovation activity and ability (Lundvall 1988). However, enterprises can only implement learning processes if they possess the ability to integrate external knowledge into their production and management activities (Ritter 1998; Koschatzky 1999). The larger the knowledge base of a firm, and the better the firm's competencies are developed to integrate external knowledge into the organisation (Le Bars *et al.* 1998: 316), the more pronounced is the ability to absorb new knowledge and to innovate. An important intermediate function is given to the so-called "Gatekeeper". Firms with a centrally organised co-ordination of external knowledge supply and knowledge distribution make themselves dependent on the absorptive capacity of their gatekeepers. Particularly in times of rapid technical change and economic transformation this dependency can be problematic (Cohen/Levinthal 1990: 132). The inclusion of several people in the firms' knowledge flow, i.e. decentralised knowledge co-ordination and distribution, reduces the risk of selective knowledge use. This allows for better access to relevant information and knowledge sources and the utilisation of them for own innovation activities (Koschatzky/Bross 1999). Thus network ability, learning and knowledge accumulation represent a cumulative process, by which firms might enter into path dependency. Firms which do not co-operate and which do not exchange knowledge reduce their knowledge base on a long-term basis and lose the ability to enter into exchange relations with other firms and organisations. On the other hand, firms which are integrated into multilayered networks, continuously improve their abilities for learning as well as their knowledge base, and concomitantly, the possibility of using new knowledge (Capello 1999).

Stressing the relevance of knowledge exchange and learning processes in the inno-vation process, evolutionary innovation economics also emphasise the importance of spatial aspects in innovation. It is argued that codified knowledge, e.g. embedded in standardised technologies, can be transferred over long distances cost, especially when the knowledge receiver is able to understand and read the code. Spatial proximity between user and producer is not necessary. Tacit knowledge, on the other hand, is only transferable through interpersonal contacts and verbal or non-verbal communication (Arnold/Thuriaux 1997: 25; Foray/Lundvall 1996: 21). Spa-tial, social, and cultural proximity is a major precondition for this transmission pro-cess. It supports the rapid exchange of information and knowledge and contributes to reducing uncertainty. Knowledge users located in such locations could benefit from this broad supply. This is especially the case in agglomerations with their di-versity in firms and the richness of research establishments (Storper 1995; Storper 1997), but also between agglomerations where proximity capital permits fast ac-cessability (Cooke 1999). On the other hand, knowledge sources are also available outside urban areas and firms are able to adopt their kind of knowledge acquisition and innovation strategies to the respective environment (Meyer-Krahmer 1985; Keeble 1997: 289). Nevertheless, the spatial environment of a firm influences its information and knowledge access and its ability for mutual learning (Kee-ble/Wilkinson 1999; Lawson/Lorenz 1999).

Regional innovation networks therefore act as catalysts in the exploitation of re-gional innovation potential (Tödtling 1994). In the ideal case, a network of local, regional and supraregional relationships arises at a regional level in the form of market-based and non-market-related interactions,[1] characterised by contacts, the exchange of information and formal and informal co-operation (cf. Figure 1-1). The purpose of these relationships for the enterprises is a reduction of their transaction costs and the efficient buying-in of complementary research capabilities. This ap-plies particularly in cases where the spectrum of potential partners' offers is known and corresponds to the enterprise's own requirements, where the number of actors exceeds a minimum threshold necessary for mutual co-operation, and where the linking of competencies promises a high degree of synergy on both sides.

Regarding the importance of innovation networks for the innovation ability of firms, the following *conclusions* can be drawn:

- Networks facilitate access to complementary, external knowledge.

- Learning processes can be implemented by co-operation and the use of external knowledge.

---

1 Market-oriented interactions are expressed as contractually agreed, financially remunerated R&D co-operation between firms, or between firms and research institutions; non-market-oriented in-teractions may arise through the exchange of ideas, the passing-on of information and the transfer of (scientific) knowledge.

- Within networks, different types of innovation-relevant knowledge and information can be exchanged. The spectrum reaches from informal information exchange to the joint execution of innovation projects.

Figure 1-1:   Regional innovation networks

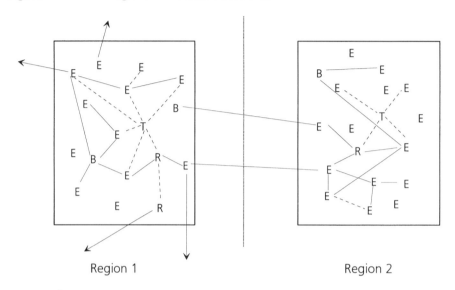

Region 1   Region 2

E: Enterprise
R: Research Institution   ——————   market-oriented interaction
T: Transfer Office   - - - - -   non-market-oriented interaction
B: Bank

Source: Koschatzky/Gundrum 1997: 205

- Besides production- and sales-based (vertical) innovation networks, horizontal networks between partners not integrated into the value chain represent an important additional source of information and knowledge.

- In addition to other production and service enterprises, research institutes are an important information and knowledge source with a bridgehead function to other networks.

- Networks contribute to development if they are open and dynamic, i.e. characterised by competition and co-operation. Closed networks can retard development.

- The absorptive capacity of an organisation influences its ability for knowledge use. The larger the already available knowledge base, the more pronounced is the ability to absorb new knowledge.

- Decentral knowledge and innovation co-ordination reduces the risk of selective knowledge use and transfer and thus supports the identification and use of relevant knowledge.

- Spatial proximity plays a role particularly in the transfer of implicit, non-codified knowledge. Nevertheless, this should not lead to the assumption that spatially limited networks are advantageous. A broad knowledge access can be realised only by a mixture of intraregional and interregional/international innovation networks.

1.3 Innovation Networking in Europe: Some Empirical Evidence

Between 1995 and 1997, a team of researchers from the Department of Economic Geography at Hanover University, the Chair for Economic Policy at Technical University Bergakademie Freiberg, the Department for Economic and Social Geography at University of Cologne, and Fraunhofer Institute for Systems and Innovation Research, Karlsruhe, has carried out an innovation survey among manufacturing firms, business-related service firms and research institutes in 11 European regions. Three very similar questionnaires were distributed in each of the regions (postal survey), especially asking about the kind and intensity of innovative (respectively research) activity and innovation networking. In total, the European Regional Innovation Survey (ERIS) collected data from more than 8,600 questionnaires, including almost 4,200 from manufacturing firms, more than 2,500 from service firms and more than 1,900 from research institutions. Response rates of these 33 surveys (three in each of the 11 regions) range from 13 % to 50 % in the subsamples (cf. Sternberg 2000 and Koschatzky/Sternberg 2000 for more details).

With regard to innovation networking, it can be concluded from the ERIS data that

- in all of the analysed regions, innovation networking plays an important role in the innovation process for the firms, although differences in intensity and spatial range of networking could be detected, depending on branch, size, technology and market orientation,

- co-operating firms are economically more successful than firms which do not co-operate, and

- spatial proximity between the co-operation partners is less important for customer-supplier relations, but plays a distinct role in horizontal networking between firms and research institutes.

Looking especially at *small firms*, an important target group in regional policy, the data analyses make it clear that

- small firms show a high preference for local and regional co-operation partners and have a much higher share of intra-regional links than large firms,

- small firms co-operate to a lesser extent with universities and other research institutes, while medium-sized and large firms make much more use of this information and knowledge pool, and

- because of their preference for local and regional partners, small firms depend greatly on the supportive quality of their regional environment. While this might not pose a problem in urban agglomerations, firms located outside urban areas might face deficits in knowledge supply which negatively affects their innovative performance.

It is therefore an important objective for regional innovation policy to fully integrate small and medium-sized firms into regional and interregional innovation networks.

1.4 Promotion of Innovation Networking by Innovation and Technology Policy

In recent years, the "network paradigm" (Cooke/Morgan 1993) has become the starting-point for policy measures aiming at the better exploitation of innovation potential, especially at the regional level. While in the years before the network idea and the possibilities for making use of spatial and cultural proximity between firms and supporting institutions were at least also implicitly applied in public promotion measures and innovation supporting services (e.g. in the Steinbeis technology transfer concept, or in the promotion of joint research projects between firms or between firms and research institutes), this paradigm is now made explicitly in innovation policy. Since these recent developments focus mainly on the region, i.e. a subnational spatial entity, as a platform for policy implementation, especially those approaches which have a regional focus are briefly described.

Before doing so, the question should be raised as to whether government, be it local, national or supra-national, has the legitimation to intervene in network-building processes which are usually organised by market forces. In general, firms depend highly on strong links with their external world. Production and marketing would not be possible without these links. If firms are unable to manage these core competencies, they will not survive. It would be market distortion if the state were to intervene at this stage. The situation looks different when it comes to the point that overall public interests are tackled. This might be the case when the market is unable to adjust according to principles of social wealth and economic prosperity. Since innovative activity not only contributes to the economic success of firms, but to the technological development, competitiveness and employment generation of a whole nation, it is the task of the government to create framework conditions which enable the economic actors to contribute to public priorities and wealth through increased innovative activity. One means for stimulating innovation is, according to theory as well as practice, the initiation, establishment and promotion of innovative links between firms and their external environment. With these objectives the government should assist those firms and institutions whose size and lack of resources, competencies and location, discriminate against them in network-building and knowledge exchange, compared to others. The assumption is that innovative potential in branches, technologies and regions which has not been utilised so far could be exploited. Here, innovation policy has a strong link to industrial, economic and regional policy. With regard to the principle of subsidiarity, the government should withdraw from network promotion when such processes could be organised by economic forces alone. Another objective of public intervention could be the improvement of the scientific and technological excellence of a nation. In this case network stimulation should not only bring those actors and competencies together which already have the potential for scientific progress and economic growth, but also those whose potential could be further exploited and synergy effects be realised by joint efforts in research and product development. Again, when the objective is fulfilled, governmental support should be concluded.

Regions are equipped with a specific permutation of production factors; these can only be considered to be optimally allocated if they are made the basis for regional technology and innovation promotion (cf. Koschatzky 1997: 185-187 for the following paragraphs). National technology policy is not usually in a position to take regional problem situations adequately into account, since neither its goals nor its instruments are adapted to coping with regional peculiarities. A policy oriented towards the optimisation of national innovation resources generally tends to increase regional disparities. Since important research institutions and industrial research laboratories are usually located in areas that are already economically favoured, public promotion of these locations tends to underline and reinforce existing regional disparities. Although financial transfers from a central government to regions in different stages of development may combat development and innovation barriers in individual instances, actions of this type will certainly cease to be suitable when measures are required which need to involve the education and training system, management competencies or information and consulting aspects. Nevertheless, regional development can also be promoted by the central government. A central regional policy aims to reduce socio-economic disparities with reference to the national average (e.g. the "Gemeinschaftsaufgabe Verbesserung der regionalen Wirtschaftsstruktur" in Germany). Infrastructural measures (involving transport, telecommunication and energy systems), regionally-differentiated investment grants and tax reductions can stimulate intraregional or external economic potential and temporarily increase the mobility of production factors oriented towards the region. In the long term, these measures can contribute to a stabilisation in the range of interregional income disparities. Particularly in the less-developed regions of Europe, measures of this kind represent a (partial) compensation for the emigration of workers and the diminishing of economic activities. A regional corrective is supplied in larger federal countries by measures at state level (e.g. the German "Länder"). Due to the relatively small volume of promotion funds, however, the extent of the impacts achieved by these Länder programmes is not generally comparable to the effects of national measures.

Through their enterprises and research institutions, regions are usually integrated into international and national technology networks. There is little scope for regionally-specific technology development, although new technologies may definitely have a regional origin. However, because of the existence of regional focuses in technology, they can be made the starting-point for an endogenous regional development strategy. An innovation- and technology-oriented regional policy is thus subject to restrictions which, although they may limit its scope for action, may also lead to promotion approaches that are sometimes very specific. When formulating measures, it is important to bear in mind that regional promotion activities are embedded in national and supranational science and technology policy, and cannot be considered in isolation from these levels. Enterprises, as well as research and development institutions, have the option chance to participate in national and supranational promotion programmes. Regional programmes must be oriented according to these higher-level promotion lines. Despite the existence of national and supranational (i.e. European) regional promotion, the amount of funding which can be made available by regional authorities for regionally-specific measures is limited.[2] Promotion measures thus have to be planned and implemented, with the involvement of regional and local actors, in a form that will meet with a high degree of acceptance and will generate a correspondingly large regional impact.

Public regional innovation promotion therefore has three tasks (Koschatzky/ Gundrum 1997: 212):

• the activation and careful complementing of regional resources for the development and application of new technologies (regional innovation conditions);

• the co-ordination and interlinking of these resources in regional innovation networks, bringing in all the relevant actors in industry, science and policy;

• the integration of these regional networks into national and international clusters of technology development and production, through the creation of active interfaces and the promotion of supra-regional co-operation.

Thus there is a need for targeted initiatives at a regional level, directed towards priority problem areas and bringing together regional players by actions that are project-based or measure-based. These initiatives have to continually secure the co-

---

[2] This aspect refers to the concept of governance and its implications for regional innovation policy, which is a constitutive element in the concept of regional innovation systems (Cooke 1998).

operation process by the provision of resources (personnel, finance, equipment) and by policy legitimation. They are mainly publicly initiated and carried out by regional key figures (promoters) in politics, industry and science, but they may be supported by external specialist advice, regional co-ordination and decision-making committees and by the promotion measures of policy at higher levels. Nevertheless, there is still room for research on the success factors of regional innovation strategies. Regarding the specific economic and social structures in regions and their different development potential, it is hard not to believe that strategies which were successful in one region could also be successful in another region. What is needed is the identification of strategy elements and the analysis of their impacts under certain, well-defined framework conditions as a kind of guideline for good practice in regional innovation promotion.

Recent examples of regionally oriented strategies are the several *initiatives of the European Commission* for the promotion of regional innovation and technology transfer by formulating and jointly implementing "Regional Innovation Strategies" (RIS) or "Regional Innovation and Technology Transfer Strategies" (RITTS). New initiatives are the "Trans Regional Innovation Projects" (TRIPS) and the RIS+ promotion (for more details about RIS projects see Landabaso/Youds 1999).

The *objectives of the RIS and RITTS exercises* are twofold (European Commission 1997):

- to improve the capacity of regional actors to develop policies which take the real needs of the business sector and the strengths and capabilities of the regional innovation system into account;

- to provide a framework within which the European Union, the Member States and the regions can optimise policy decisions regarding future investments in research and technological development, innovation and technology transfer initiatives at regional level.

By definition, the active participation of the key stakeholders of the region is required in order to validate the entire exercise from the definition of a work programme to testing the feasibility of pilot projects. This means that regional networks should be built, in order to establish links between all relevant regional innovation actors. A summary of similarities and differences between RITTS and RIS projects is shown in Table 1-1 (European Commission 1997).

Table 1-1:   Characteristics of RITTS and RIS projects

RITTS - Regional Innovation and Technology Transfer Strategies and Infrastructures	RIS - Regional Innovation Strategies
Projects designed to evaluate, develop and optimise regional infrastructure and policies and strategies for supporting innovation and technology transfer	Projects designed to create partnerships among key actors in a region with a view to defining an innovation strategy for the region in the context of regional development policy
Projects are self-standing and are carried out for their intrinsic benefits to the region	Projects should seek to improve the effectiveness with which EU Structural Funds are used for promoting innovation schemes
A project may involve only part of a region (no formal administrative structure)	A RIS must cover a NUTS level II classified region
The project beneficiary is not necessarily a regional authority (e.g. an innovation agency, university, etc.)	The project beneficiary must be the authority responsible for the economic development of the region
Projects may be financed throughout the EU and the European Economic Area	Confined to regions where a significant share of the population is an ERDF-assisted area

In *Germany*, the federal government has recently paid great attention to the regions in the implementation of technology policy measures.[3] A first successful attempt was made in 1996 by initiating the *BioRegio contest*. Its major objective was to stimulate firm foundations and the location of foreign biotechnology companies in Germany, to accelerate growth in existing biotechnology enterprises and to ensure the supply of sufficient seed and venture capital to improve the competitive situation of Germany in biotechnology. In a competition procedure three regions with appropriate research potential were selected: Munich, the Rhine-Neckar Triangle (Heidelberg, Ludwigshafen, Mannheim) and the Rhineland (Cologne, Aachen, Düsseldorf, Wuppertal), each will be subsidised with 50 million DM by the Federal Ministry for Education and Research (BMBF) until 2001. The regions and the advantage of spatial proximity between industry, research and venture capitalists were made the starting point for network generation in and between the regions and international biotechnology research, testing and production. Besides the fairly vague

---

3   As a consequence, more information and data are needed on R&D and innovation activities at the regional level. Although a broad spectrum of indicators is available for the measurement of the technological and scientific performance of a nation, the measurement of regional R&D and innovation processes and the availability of respective data still has to be improved. For this reason the department "Innovation Services and Regional Development" of Fraunhofer ISI carried out a project on behalf of the Federal Ministry for Education and Research (BMBF) together with Deutsches Institut für Wirtschaftsforschung, Berlin, Institut für Weltwirtschaft, Kiel and Niedersächsisches Institut für Wirtschaftsforschung, Hanover, titled "Regional distribution of innovation and technology potential in Germany and Europe". It aimed at an analysis of regional innovation strategies as well as the elaboration and application of a set of suitable indicators for measuring regional innovation activities (cf. BMBF 2000a for more details).

concept of "competence centres", the Bioregio contest opened the door for other national activities explicitly implemented at a regional level.

The two most important initiatives are the EXIST contest and the InnoRegio contest. It is the general objective of the *EXIST contest* that concepts for regional co-operation between universities, polytechnics, the business sector and further partners should be initiated. Four guidance objectives are pursued:

(1) The permanent establishment of a "Culture of Entrepreneurship" in teaching, research and administration at the universities.

(2) The consistent transfer of scientific research results to economic value-added.

(3) The specific promotion of the huge potential of business ideas and founder personalities at universities.

(4) The increase in firm foundations and innovative services, resulting in appropriate labour market effects.

Exemplary regional networks are to be created, which should serve as good practice models for the efficient support of firm foundations from universities and research institutes by bundling regional forces, through optimised structures and innovative strategies. Five winner regions were selected by a jury, namely: Dresden ("Dresden exists"), Ilmenau/Jena/Schmalkalden ("GET UP"), Karlsruhe/Pforzheim ("KEIM"), Stuttgart ("PUSH!"), and Wuppertal/Hagen ("bizeps"). The EXIST programme will be given 30 million DM per year by the federal government (BMBF 2000b).[4]

The last regional activity to be mentioned is the *InnoRegio contest* which is confined to the new federal states. Its main objective is the sustainable improvement of the employment situation in the new federal states and the strengthening of their competitive ability (cf. Koschatzky/Zenker 1999). In order to achieve this objective, concepts and projects are to be developed on the regional level aiming at the utilisation of innovation potential. For initiating networking activities, innovation dialogues are started up. The establishment of regional networks should be fostered, in which people from different fields of activity should engage in joint innovation and learning projects. They are expected to develop ideas and visions in new co-operation beyond administrative borders or department barriers. It is assumed that regional networks bring the creativity, competence and motivation of different actors from research, business and society together in a new way. These networks are expected to be the basis for developing a competitive research, education and economic profile of the region and thereby create new employment and market opportunities.

---

4   The scientific evaluation of the five regional networks is being carried out by the Department "Innovation Services and Regional Development" of Fraunhofer ISI.

The first phase of the contest started in April 1999 with the call for proposals. Up to the deadline on 15 August 1999, more than 440 proposals had been delivered which showed a great variety of ideas. They ranged from the special use of modern information technologies, to future-oriented education projects and specific forms of "soft tourism". From these applications the jury pre-selected 50 concepts from which 25 qualified as "InnoRegios". The central evaluation criteria of the jury were the novelty of the approach, the persuasive power of regional co-operation and the expected benefit for the region. The 25 regions will be supported with 300,000 DM each by BMBF for elaborating their strategies for regional development until summer 2000. Afterwards the implementation of these strategies will also be financially supported (BMBF 1999).

Although all the programmes and contests mentioned are based on network-building, they aim at different objectives. In the cases of RITTS/RIS and InnoRegio, network formation is used as an instrument for bringing people together and for formulating a joint regional innovation strategy. In this way, bottlenecks and economic problems should be overcome and employment should be secured. Innovative is all that is new to the region, so that innovation is not confined to technological progress. These strategies should also give development chances to regions which so far only have less developed innovation and economic potential. Another strategy is employed by the BioRegio and the EXIST contest. Both are mainly aimed at improving the technological and economical competitive advantage of Germany. In the case of BioRegio, scientific excellence in biotechnology research is linked with industrial needs and interests, so that not only can the share of new products reaching the market stage increase, but also so that new firms for biotechnological product and process development can emerge. EXIST, on the other hand, aims at improving the entrepreneurial climate in Germany in general and at universities, polytechnics and research institutes in particular by transferring good practice strategies, developed in five model regional networks, to other regions and by stimulating similar activities in other locations.

1.5   Structure and Objectives of this Volume

For several years, the department "Innovation Services and Regional Development" (IR) of the Fraunhofer Institute for Systems and Innovation Research (FhG-ISI) has been working on the topics presented in this reader. The focal point of analysis is in the fields of technology and innovation-oriented regional research, innovation policies and technology transfer in Central and Eastern Europe, as well as innovation financing and further services beneficial for technologically oriented companies. Numerous projects have been carried out for public and private clients as well as research foundations. Many of them deal with firms and supporting services for innovation. Among many others the scientific monitoring of three important pilot schemes of the BMBF, i.e. the "Promotion of New Technology-Based Firms" (TOU), "Business Investment Capital for New Technology-Based Firms" (BJTU) and "Promotion of New Technology-Based Firms in the New Federal States" (TOU-NBL), the piloting of a "Network for Technological Expertise" (NTG) for the German savings banks, and the piloting of a business angels network called "Business Angels Venture (BAV) for Deutsche Bank AG and other partners should be mentioned as examples – among many others – of this kind of research activity.

As a major resource for the support of innovation processes, networks are therefore a crucial factor of the scientific-theoretical work and the consultancy of the department. After successfully completing a first IR conference on "Financing of SMEs" in 1998, a decision was made to focus the 1999 conference on innovation networking. This topic was the subject of the introductory lectures, which aimed to define and elaborate the scope and economic and political importance of innovation networking, and of four workshops planned. This volume summarises the scientific discussion of the conference by presenting a selection of papers which address innovation networking from theoretical and political viewpoints.

The book is organised into four sections. Besides this contribution, the first section (*"Introduction to the Subject"*) contains a second introductory paper by *Michael Fritsch*. He will shed some light on the economic perspective of networking. According to his analysis, a central feature of networking is a relatively high degree of division of labour. A further important characteristic of networks is the redundancy of vertical relationships which leads to a number of advantages for the members of the network and to the positive performance of the network as a whole. Fritsch comes to the conclusion that horizontal co-operation may play a role in networks but is probably not as important as vertical relationships. Finally, some alternative measures aimed at stimulating the development of innovation networks are discussed.

Section II deals with *"Knowledge and Learning in Innovation Networks"*. Small and medium-sized enterprises, in particular, do not possess the resources for acquiring the knowledge necessary for complex innovation and are therefore dependent on co-operation with other actors in a network. Production and transfer of knowledge, its application in innovation processes as well as the advancement of knowledge requires a functioning network of different partners to support region-specific strengths in knowledge management and innovation. Crucial for regional innovation success are thus not only technological knowledge, but also an efficient knowledge exchange and the advancement of the existing knowledge base. A region, defined as a network of different actors, can acquire and develop specific competencies, leading to competitive advantages and economic success. Against this background, four papers deal with different aspects of knowledge exchange and learning at the regional level. *Emmanuel Muller* deals with theoretical aspects of the interrelations between knowledge, innovation and regions and especially focuses on the interactions between industrial SMEs and knowledge-intensive business services (KIBS). The latter group of firms and the role they play in innovation processes is addressed in more detail in the contribution by *Simone Strambach*. She elaborates on the strategic significance of KIBS in innovation systems and argues that innovation policy should promote interaction and learning processes between the demand and the supply side. Institutions of technological infrastructure are responsible for the management and advancement of the knowledge base, for the support of interactions between enterprises and for the supply of expert knowledge. Their role in knowledge generation and diffusion is discussed in the paper by *Antoine Bureth and Jean-Alain Héraud*. Finally, *Javier Revilla Diez* presents results from the European Regional Innovation Survey with respect to the importance of innovative links for firm start-ups and especially with respect to research institutes in the metropolitan regions of Stockholm, Vienna and Barcelona.

Section III deals with *"Innovation Networks in Transition"*. The starting point here is that a series of Central and Eastern European Countries are negotiating their accession to the EU. Within the last years, technologically outdated state enterprises have become competent partners in trade, production and research and development. Now own potential should be utilised by innovation and networking and by integration into global networks. In the first case success stories of innovative regions and industries in western industrialised countries (e.g. the innovation orientation in Baden-Württemberg or in the Silicon Valley) might be transferred to Eastern Europe. In Western Europe, small and medium-sized private enterprises are the backbone of a flexible adjustment to market trends and of innovative growth. Cooperation among themselves proved to be a success factor in order to adjust to size disadvantages with respect to large companies. Regional innovation networks can improve the competitive situation, but they can also preserve the status quo and prevent necessary adjustments when networks are based on old ties. Furthermore, partnerships with foreign enterprises can offer access to technologies and management know-how. Another important issue in this respect is the access to capital for innovation financing. In the first contribution to this section, *Günter H. Walter* discusses the network approach as a suitable model for innovation policy in CEECs, taking Slovenia as an example for further elaboration of his thoughts. The kind and intensity of innovation networking between industry and research institutes from the viewpoint of both groups is analysed in the paper by *Knut Koschatzky and Ulrike Bross*. They present an empirical study based on data of the European Regional Innovation Survey collected in Slovenia. *Slavo Radosevic* provides the basis for a better understanding of the integration through industrial networks in wider Europe. He assumes that the integration of CEECs into the EU is only possible if production and technology integration reinforce market integration, otherwise CEECs will be isolated in terms of production and technology links and will depend excessively on budgetary transfers. The last two chapters of this section deal with German experiences. *Franz Pleschak and Frank Stummer* describe the situation of industrial research in East Germany and develop strategies for the improvement of innovative activity and economic competitiveness in East German companies by the establishment of innovative networks. *Marianne Kulicke* gives an impression of the role of regional venture capital companies in Germany and their integration into innovation networks and draws conclusions from the German experience in innovation finance which are of relevance for transition countries.

Finally, Section IV deals with *"Innovation Networks and Regional Innovation Policy"*. Here, specific aspects briefly addressed in this introductory paper will be elaborated on in more detail. Based on the concept of regional innovation systems, *Andrea Zenker* offers an analytical basis for discussing concepts, strategies and success factors of network-based regional innovation strategies. She not only describes the conceptual elements of these strategies, but also illustrates her analysis with examples from practice. *Herbert Berteit* directs our view a second time to East Germany by discussing the framework conditions for innovation networks in the new Federal States, by presenting some statistical indicators for innovation networks and by deriving perspectives of economic and structural policy for the promotion of innovation networks in East Germany. In the final paper of this section and in this volume, *Mikel Landabaso, Christine Oughton and Kevin Morgan* deal with innovation networks and regional policy in Europe. They draw on the concepts of regional innovation systems and learning regions for pointing to the importance of integrating and interlinking all actors of a region to their regional innovation system. Experiences based on the RIS activities of the European Commission are presented, and conclusions for European innovation and regional policy are drawn.

To sum up, it is the objective of this volume to discuss different facets of innovation networking by presenting theoretical, empirical and policy-oriented contributions based on European experiences. During the conference, the following questions served as guiding principles for discussion in the different workshops and in the final panel discussion:

- What characteristics do successful innovation networks display in different national and regional contexts?

- What political recommendations can be given for the initiation and support of innovation networks?

- How can innovation networks be designed in a transformation context? How can they function best?

- What results from Western European regions are applicable for Central and Eastern European transformation economies?

- What significance do knowledge and collective learning processes have in the innovation context?

- Which results are available concerning the importance of knowledge within companies' innovation processes?

These questions should also serve as guiding principles for reading and should help the reader to discover whether the papers in this volume have contributed to an improved insight into the importance of innovation networks for improved economic and technological competitiveness and for regional development.

1.6 References

ARNOLD, E./THURIAUX, B. (1997): *Supporting Companies' Technological Capabilities*. Brighton: Technopolis Ltd.

BMBF [BUNDESMINISTERIUM FÜR BILDUNG UND FORSCHUNG] (1998): *EXIST. Existenzgründer aus Hochschulen. 12 regionale Netzwerke für innovative Unternehmensgründungen*. Bonn: BMBF.

BMBF [BUNDESMINISTERIUM FÜR BILDUNG UND FORSCHUNG] (1999): *InnoRegio. Innovative Impulse für die Region. Ausschreibungsbroschüre*. Bonn: BMBF.

BMBF [BUNDESMINISTERIUM FÜR BILDUNG UND FORSCHUNG] (Ed.) (2000a): *Zur technologischen Leistungsfähigkeit Deutschlands. Zusammenfassender Endbericht 1999*. Bonn: BMBF.

BMBF [BUNDESMINISTERIUM FÜR BILDUNG UND FORSCHUNG] (2000b): EXIST - University-based start-ups. Networks for innovative company start-ups. Bonn: BMBF.

CAPELLO, R. (1999): Spatial Transfer of Knowledge in High Technology Milieux: Learning Versus Collective Learning Processes, *Regional Studies*, 33, pp. 353-365.

CASTI, J.L. (1995): The Theory of Networks, BATTEN, D./CASTI, J./THORD, R. (Eds.) *Networks in Action. Communication, Economics and Human Knowledge*. Berlin: Springer, pp. 3-24.

CIMOLI, M./DOSI, G. (1996): Technological paradigms, patterns of learning and development: an introductory roadmap, DOPFER, K. (Ed.) *The Global Dimension of Economic Evolution. Knowledge Variety and Diffusion in Economic Growth and Development*. Heidelberg: Physica-Verlag, pp. 63-88.

COHEN, W./LEVINTHAL, D. A. (1990): Absorptive capacity. A new perspective on learning and innovation, *Administrative Science Quarterly*, 35, pp.128-152.

COOKE, P. (1998): Introduction. Origins of the concept, BRACZYK, H.-J./COOKE, P./HEIDENREICH, M./KRAUSS, G. (Eds.) *Regional Innovation Systems. The role of governance in a globalized world*. London: UCL Press, pp. 2-25.

COOKE, P. (1999): *Regional Innovation Systems: General Findings and Some new Evidence from Biotechnology Clusters*. Paper prepared for NECTS/RICTES Conference "Regional Innovation Systems in Europe", 30 September - 2 October, 1999, Donostia/San Sebastian.

COOKE, P./MORGAN, K. (1993): The Network Paradigm. New Departures in Corporate and Regional Development, *Society and Space*, 11, pp. 543-564.

COOMBS, R./HULL, R. (1998): 'Knowledge management practices' and path-dependency in innovation, *Research Policy*, 27, pp. 237-253.

COWAN, R./FORAY, D. (1997): The Economics of Codification and the Diffusion of Knowledge, *Industrial and Corporate Change*, 6, pp. 595-622.

DEBRESSON, C./AMESSE, F. (1991): Networks of innovators. A review and introduction to the issue, *Research Policy*, 20, pp. 363-379.

EUROPEAN COMMISSION (1997): RIS/RITTS Guide. Brussels.

FORAY, D./LUNDVALL, B.-Å. (1996): The Knowledge-Based Economy: from the Economics of Knowledge to the Learning Economy, ORGANISATION FOR ECONOMIC CO-OPERATION AND DEVELOPMENT (Ed.) Employment and Growth in the Knowledge-based Economy. Paris: OECD, pp. 11-32.

FREEMAN, C. (1991): Networks of innovators: A synthesis of research issues, *Research Policy*, 20, pp. 499-514.

FRITSCH, M. (1992): Unternehmens-"Netzwerke" im Lichte der Institutionenökonomik, BÖTTCHER, E./HERDER-DORNEICH, PH./SCHENK, K.-E./SCHMIDTCHEN, D. (Hrsg.) *Jahrbuch für Neue Politische Ökonomie. 11. Band: Ökonomische Systeme und ihre Dynamik*. Tübingen: J.C.B. Mohr, pp. 89-102.

FRITSCH, M./KOSCHATZKY, K./SCHÄTZL, L./STERNBERG, R. (1998): Regionale Innovationspotentiale und innovative Netzwerke, *Raumforschung und Raumordnung*, 56, pp. 243-252.

GRABHER, G. (1993): The weakness of strong ties: the lock-in of regional development in the Ruhr area, GRABHER, G. (Ed.) *The embedded firm. On the socio-economics of industrial networks*. London: Routledge, pp. 255-277.

GRANOVETTER, M. (1982): The Strength of Weak Ties. A Network Theory Revisited, MARSDEN, P. V./LIN, N. (Eds.) *Social Structure and Network Analysis*. Beverly Hills: Sage, pp. 105-130.

KARLSSON, C./WESTIN, L. (1994): Patterns of a Network Economy – An Introduction, JOHANSSON, B./KARLSSON, C./WESTIN, L. (Eds.) *Patterns of a Network Economy*. Berlin: Springer, pp. 1-12.

KEEBLE D. (1997): Small Firms, Innovation and Regional Development in Britain in the 1990s, *Regional Studies,* 31, pp. 281-293.

KEEBLE, D./WILKINSON, F. (1999): Collective Learning and Knowledge Development in the Evolution of Regional Clusters of High Technology SMEs in Europe, *Regional Studies*, 33, pp. 295-303.

KOSCHATZKY, K. (1997): Innovative Regional Development Concepts and Technology-Based Firms, KOSCHATZKY, K. (Ed.) *Technology-Based Firms in the Innovation Process. Management, Financing and Regional Networks*. Heidelberg: Physica-Verlag, pp. 177-201.

KOSCHATZKY, K. (1999): Innovation Networks of Industry and Business-Related Services – Relations between Innovation Intensity of Firms and Regional Inter-Firm Co-operation, *European Planning Studies*, 7, pp. 737-757.

KOSCHATZKY, K./BROSS, U. (1999): *Struktur und Dynamik von regionalen Innovationsnetzwerken unter Transformationsbedingungen – das Beispiel Slowenien.* Karlsruhe: Fraunhofer ISI, Arbeitspapier Regionalforschung, No. 20.

KOSCHATZKY, K./GUNDRUM, U. (1997): Innovation Networks for Small Enterprises, KOSCHATZKY, K. (Ed.) *Technology-Based Firms in the Innovation Process. Management, Financing and Regional Networks.* Heidelberg: Physica-Verlag, pp. 203-224.

KOSCHATZKY, K./STERNBERG, R. (2000): R&D co-operation in innovation systems – Some lessons from the European Regional Innovation Survey (ERIS), *European Planning Studies,* 8, pp. 487-501.

KOSCHATZKY, K./ZENKER, A. (1999): *Innovative Regionen in Ostdeutschland – Merkmale, Defizite, Potentiale.* Karlsruhe: Fraunhofer ISI, Arbeitspapier Regionalforschung No. 17.

LANDABASO, M./YOUDS, R. (1999): Regional Innovation Strategies (RIS): the development of a regional innovation capacity, *SIR-Mitteilungen und Berichte,* Bd. 17, pp. 1-14.

LAWSON, C./LORENZ, E. (1999): Collective Learning, Tacit Knowledge and Regional Innovative Capacity, *Regional Studies,* 33, pp. 305-317.

LE BARS, A./MANGEMATIN, V./NESTA, L. (1998): Innovation in SMEs: The Missing Link, HIGH-TECHNOLOGY SMALL FIRMS CONFERENCE *The 6th Annual International Conference at the University of Twente, the Netherlands. Proceedings,* 1. Twente: University of Twente, pp. 307-324.

LUNDVALL, B.-Å. (1988): Innovation as an interactive process: From user-producer interaction to the national system of innovation, DOSI, G./FREEMAN, CH./NELSON, R./SILVERBERG, G./SOETE, L. (Eds.) *Technical Change and Economic Theory.* London: Pinter, pp. 349-369.

MEYER-KRAHMER, F. (1985): Innovation Behaviour and Regional Indigenous Potential, *Regional Studies,* 19, pp. 523-534.

NONAKA, I./TAKEUCHI, H. (1995): *The Knowledge-Creating Company. How Japanese Companies Create the Dynamics of Innovation.* New York: Oxford University Press.

POWELL, W.W. (1990): Neither market nor hierarchy, STAW, B.M./CUMMINGS, L.L. (Eds.) *Research in Organizational Behaviour,* 12. Greenwich: JAI Press, pp. 295-336.

REID, S./GARNSEY, E. (1998): How Do Small Companies Learn? Organisational Learning & Knowledge Management in the High-Tech Small Firm, HIGH-TECHNOLOGY SMALL FIRMS CONFERENCE *The 6th Annual International Conference at the University of Twente, the Netherlands. Proceedings,* 1. Twente: University of Twente, pp. 391-401.

RITTER, T. (1998): *Innovationserfolg durch Netzwerkkompetenz. Effektives Management von Unternehmensnetzwerken.* Wiesbaden: Gabler Verlag.

SEMLINGER, K. (1998): *Innovationsnetzwerke. Kooperation von Kleinbetrieben, Jungunternehmen und kollektiven Akteuren.* Eschborn: RKW.

SPENDER, J.-C. (1996): Making Knowledge the Basis of a Dynamic Theory of the Firm, *Strategic Management Journal*, 17, pp. 45-62.

STERNBERG, R. (2000): Innovation Networks and Regional Development – Evidence from the European Regional Innovation Survey (ERIS): Theoretical Concepts, Methodological Approach, Empirical Basis and Introduction to the Theme Issue, *European Planning Studies*, 8, No. 4.

STORPER, M. (1995): The Resurgence of Regional Economies, Ten Years Later: The Region as a Nexus of Untraded Interdependencies, *European Urban and Regional Studies,* 2, pp. 191-221.

STORPER, M. (1997): *The Regional World. Territorial Development in a Global Economy.* Guilford Press, New York.

TÖDTLING, F. (1994): Regional networks of high-technology firms – the case of the Greater Boston area, *Technovation*, 14, pp. 323-343.

WILLIAMSON, O.E. (1975): *Markets and Hierarchies: Analysis and Antitrust Implications. A Study in the Economics of Internal Organization.* New York, London: The Free Press/Macmillan.

WILLIAMSON, O.E. (1985): *The Economic Institutions of Capitalism. Firms, Markets, Relational Contracting.* New York, London: The Free Press/Macmillan.

2. Innovation by Networking: An Economic Perspective

Michael Fritsch

632

L22

2.1 Introduction

The concept of innovation "networks" has by and large been developed in social sciences other than economics. A whole number of quite different definitions of what a network is can be found in the literature (see the contribution of Knut Koschatzky in this volume for an overview). According to this heterogeneity of concepts and definitions, a considerable divergence of hypotheses concerning the main forces that govern such networks exist. In this contribution, I want to outline an economic approach to the analysis of innovation networks. That is, to show why it might be advantageous for the innovation activities of individual economic actors to be embedded in a network of relationships to other actors. My argument will focus on those kinds of networks that mainly consist of private sector firms. Pure policy networks are, therefore, excluded.

Seen from an economic perspective, networking represents a means to improve the efficiency of innovation processes and, in particular, to overcome some impediments to a division of innovative labour. Therefore, the specific problems of labour division in the field of innovation constitute the starting point of the analysis here (Section 2.2). Based on that, I will give a definition of what a network is (Section 2.3) and then explicate the potential advantages of network relationships for the firms involved (Section 2.4). Section 2.5 contains some final remarks for policy and for further research.

2.2 Impediments to a Division of Innovative Labour

Division of labour in the field of innovation is characterised by a number of *specific problems*. These problems are a consequence of the very nature of innovation processes as well as of certain characteristics of information as a subject of market transaction. Because the combination of existing knowledge and the generation of new knowledge constitutes the core of innovation activities, a transfer of relevant information plays an essential role in interaction on R&D.

One main impediment to a division of innovative labour is caused by the fact that the output of innovation activities cannot be completely specified beforehand in a relevant contract. This implies that the resulting incomplete contracts leave room for *opportunistic behaviour* of contractual parties, i.e., self-serving interpretation of the terms of the contract to the disadvantage of other contract parties. Due to this danger of opportunistic behaviour, economic actors may avoid contracting out certain tasks of the innovation process that otherwise would be purchased from external sources.

Another problem for a division of innovative labour is *asymmetric information* that may constitute a severe impediment for trading information on markets. The reason is that in order to make suppositions about the economic value of certain information, one needs to know its properties. Therefore, the supplier of information should describe the characteristics of that information and, in many cases, this implies a more or less complete disclosure. However, once a potential customer possesses the information, he has no reason to purchase it. Therefore, information that is intended to be sold cannot be completely disclosed. Consequently, the supplier has better knowledge of the subject of the potential market transaction than the customer. This may hamper the trading of information on a market and, thus, interaction on R&D.

A third possible difficulty concerns the *transfer of information* as such. One obstacle to information transfer may be, for example, that the information is "tacit", i.e., not completely codifiable so that it can only be communicated face-to-face or by a transfer of the person who possesses the knowledge. Moreover, the identification and the use of relevant information may require a certain "absorptive capacity" (Cohen/Levinthal 1989; for a comprehensive treatment of problems of information transfer see v. Hippel 1994). Co-operation with regard to innovation may also be hampered by the danger of *uncontrolled knowledge flows*, i.e., that by co-operating on R&D the transaction partner comes into possession of sensitive information without paying an adequate compensation for it.

A further problem of a division of innovative labour may be due to the fact that R&D processes often require very special inputs that are not commonly traded in large markets. This *rareness of appropriate inputs* in many cases is a result of the very nature of innovation activities characterised as generating something new: new products or processes may require new or very specialised inputs that are not readily available. In fact, markets for inputs to R&D processes may be fairly "thin" with only very few suppliers able to fulfil the desired task and transactions taking place rather infrequently. Because suppliers are rare, an immense amount of effort may be required in terms of *search costs* to identify a suitable transaction partner for a division of innovative labour. Since such transaction-specific investment will be "sunk" if the respective relationship is abandoned, there is an incentive to utilise such relationships – once established – over a longer period of time in order to reap the rewards from this investment. Moreover, if only few transactions take place, a clear market price may not exist so that negotiations about the price and further conditions of an exchange tend to be rather costly.

As a result of these problems, many contributions to innovation processes cannot be easily traded on anonymous "spot markets". Therefore, a division of innovative labour between different organisations may require incompletely specified, long-term agreements ("relational contracting") that imply a considerable degree of trust and co-operative spirit.[1]

2.3   What is a "Network"?

For the present purpose any set of social relationships may be called a "network" if it consists of at least three individuals or institutions and is characterised by some redundant vertical relationships that are only incompletely specified. The networks under inspection here consist mainly of private firms and correspond to the concept of "industrial districts". Redundancy of business relations means that there is a tendency for customers to have more than one supplier of certain goods and that suppliers are not dependent on only one customer. Many of the network relationships are characterised by a long-term orientation. One possible reason for such a long-term orientation is that establishing a relationship with a certain transaction partner may require high relation-specific investment that will be "sunk" if the respective relationship is abandoned. Moreover, repeated transactions over a longer time-period create conditions conducive to the emergence of trust and reputation that are needed to overcome the problems caused by asymmetric information and incompletely specified contracts. Due to the considerable incentives to be gained from relation-specific investments over a longer time period, networks are characterised by a pronounced tendency to solve conflicts by dispute ("voice") because the "exit"-alternative, i.e., abandoning the relationship, would be rather costly. In many cases, networks are characterised by a certain network-"culture", i.e., a set of shared values or certain modes of conduct for transactions and for conflict-solving. Economically, such a culture constitutes a means of reducing uncertainty about the behaviour of other members of the network (cf. Carr/Landa 1983). Although network relationships tend to be long term in character they need not be very tight. In many cases, such relationships may be described as "weak ties" and "loose coupling" (Granovetter 1973; Weick 1976).

---

1   See MacNeil (1978) for a detailed characterisation of the different types of agreements.

Another remarkable feature of many networks that can be found in reality is the clustering in space, i.e., that economic activity in a certain technological field - particularly R&D-activity - is concentrated in particular regions (Porter 1998). This may indicate different things. First, clustering can be caused by the presence of inputs (e.g., a particular university department, a differentiated labour market) in a region that is important for a certain type of innovation activity. Second, there may be considerable positive technological externalities (e.g., knowledge spillovers from R&D) at work that are limited to a region. And third, clustering can be caused by the advantages or the need for being located in spatial proximity when collaborating on R&D. This holds particularly for innovation activities where frequent face-to-face contacts are of considerable importance.

2.4   Advantages of Network Relationships

My main hypothesis with regard to networks of private firms as they are described in the literature is that this form of interaction allows firms to realise a relatively high degree of labour division. The analysis will clearly show that this mainly concerns vertical relationships which are regarded in Section 2.4.1. Further advantages may emerge from horizontal relationships in networks (see Section 2.4.2). In the literature, a number of additional arguments can be found that assert some general benefits of network-type relationships, be they vertical or horizontal. These aspects are explicated in Section 2.4.3.

2.4.1   Benefits of Vertical Disintegration and Redundancy of Relationships

One important feature of private firm networks or industrial districts often described in the literature is a relatively high degree of division of labour (cf. the contributions in Camagni 1991 and in Pyke/Becattini/Sengenberger 1990). Many firms involved in such networks concentrate exclusively on only a few specific steps of the production process and leave the rest to other actors that are specialised to a similar degree. Such a high level of labour division may bring a number of benefits for the individual economic actor as well as for the network as a whole. One of the advantages on the level of the individual member of the network could be an increase of flexibility due to reduced complexity (lower internal transaction costs) of a "lean" organisation in which less tasks are performed.

Other benefits of an increased division of labour in networks presuppose some degree of redundancy in relationships with customers or suppliers. Obviously, redundancy of vertical relationships represents a central issue of a division of labour within networks. For the individual firm, the *benefits of an increased division* of labour together with redundant relationships encompass a number of issues:

- Higher productivity or quality of output due to a higher degree of specialisation. This effect is mainly caused by the higher volume of production realised by actors who perform the relevant task not only for internal demand but also for other actors.

- Better "match" of available inputs if there is choice between different supplies. As a consequence of a better match, less effort is necessary to adapt the respective goods or services to a firms' specific needs. For a better "match" of inputs to occur it is necessary that there are several suppliers available (= potential redundancy in supplier relations) and that the inputs offered by these different sources are characterised by some degree of heterogeneity. The more differentiated the supply, the higher the probability of finding a perfect fit.

- Better chances of avoiding bottlenecks if one supplier is unable or unwilling to deliver certain goods or to perform a certain task. Therefore, internal or external bottlenecks may be overcome much more smoothly than in the case of a complete internal provision or a single sourcing situation, i.e., when a relationship to only one supplier exists.

- Opportunity to compare the cost and quality of different suppliers' goods and services. Such benchmarking may not be limited to the cost and quality of a product but also to the respective production processes leading to a faster diffusion of process innovation. If, for example, a certain supplier has successfully implemented a process innovation the customer may be interested in having his other suppliers adopt this product innovation and, therefore, engage in actively stimulating the diffusion of the innovation.

In addition to the positive effects of outsourcing and division of labour in innovation networks there are a number of further potential benefits of redundancies. Such *advantages of redundant relationships* are:

- A relatively fast diffusion of information and innovation within networks. Fast diffusion is mainly the result of redundancy of vertical relationships and of relatively rich information flows within incompletely specified interaction. If a supplier is not dependent on a certain customer but has also established relationships with other customers, one may be able to benefit from innovations that have been developed in the interaction of the supplier with the other customers.

- Stability and safeguarding of relation-specific investments. If a supplier delivers a large fraction of his output to one particular customer, his existence may be endangered if the customer breaks off the relationship for any length of time. However, if such a dependent supplier is forced to exit the market due to a loss in demand, the customer cannot go back to his former supplier and all relation-specific investments will be sunk. Non-dependency on a specific customer makes the supplier more resistant to a temporary break-off of the relationship and enables both parties to benefit from the relation-specific investment over a longer period of time.

- Automatic correction of errors or misperceptions of transferred information. Redundancy of relationships implies that certain information will be communicated from different sources. Therefore, false or misperceived information can be corrected more or less "automatically".

- Limiting the scope for exploitation. Redundancy of vertical relationships implies a possibility of bypassing certain transaction partners and, therefore, limits the danger of being exploited that is always present when some degree of dependency exists.

- Generation of variety. Redundancy may be seen as a necessary but not sufficient condition for the emergence of different approaches and solutions. If, for example, a number of firms try to match a certain demand, their approaches may differ considerably. Variety is a prerequisite for a better match of inputs to occur and may, therefore have welfare-increasing effects in the static sense. Variety may be even more important for the dynamic performance of the system. The more different variants available, the higher the probability of finding solutions to unforeseen problems that may arise in the future.

It has already been mentioned that redundancy of relationships implies the availability of alternatives, and the more acceptable alternatives there are, the lower the danger or probability of being dependent on a certain exchange partner. Nondependency due to redundant relationships is an issue that has been characterised in the literature as "loose ties" (Grabher 1993; Granovetter 1973, 1982; Weick 1976). Loose coupling of different elements or organisations may have a number of consequences for the dynamic properties of the system as a whole. Most of the effects of loose ties as mentioned in the literature – such as higher flexibility, higher variety, relatively fast diffusion of innovations – have already been discussed in the context of redundancy of relationships (Section 2.4.2). However, a further possible effect of weak ties is not so obviously a result of redundancy, and this is a relatively high sensitivity of loosely coupled systems with regard to their socio-economic environment. This hypothesis proposed by Mark Granovetter (1973) asserts that weak ties are more likely to link members of different groups than strong ones. Granovetter argues that strong ties tend to be concentrated within a particular group with the result that the members of such a group possess approximately the same set of information. Accordingly, the highest probability of learning something new is related to communication across greater social distance because the weak ties between such distant groups allow for more heterogeneity of information among communication partners than that available when the partners are connected by strong ties.

Despite the obvious benefits of outsourcing and redundancy, it should not be overlooked that establishing co-operative relationships with a number of more or less similar partners implies multiple relation-specific investment and may, therefore, be rather costly. For this reason, the number of redundant relationships will be limited. From an economic perspective, there is an optimum in the number of redundant relationships that results from a cost-benefit comparison.

2.4.2 Possible Effects of Horizontal Network Relationships

Horizontal relationships in networks have two types of advantages. The first type of advantage is the pooling of R&D resources for a specific innovation. As far as face-to-face contacts are necessary for conducting such an R&D-co-operation, spatial proximity may be conducive to establishing and maintaining such relationships. Empirical research suggests that a relatively weak and informal type of horizontal R&D-co-operation, so-called information trading, may be of particular importance for innovation networks (cf. Saxenian 1994). Information trading denotes an exchange of information, in most cases technical knowledge, between personnel of competing firms (von Hippel 1987). Often, a major aim of horizontal co-operation is to overcome disadvantages due to small size caused by indivisibilities of processes or resources. Another possible objective that may be relevant in many cases is spreading the costs of R&D among those firms that will probably benefit from the result. This motive of horizontal R&D co-operation may be of particular importance when the flow of knowledge cannot be readily controlled (cf. Katz/Ordover 1990).

A second possible benefit of horizontal networking is the joint use of certain inputs, such as infrastructure facilities or factor markets. Joint use of inputs or of input markets may also work as a vehicle for knowledge "spillovers". This can be the case particularly in regard to the labour market due to the knowledge flows which occur when personnel shift between different employers. The availability of such inputs or markets may also constitute an important location factor for firms or industry of a certain type and attract start-ups or relocations into the area.

2.4.3 General Characteristics of Network Relationships

The characterisation of horizontal and vertical interaction in networks makes clear that a relatively high share of these relationships is long term and only incompletely specified. Several authors emphasise that such kinds of relationship tend to be characterised by a high degree of openness among the interacting parties and by relatively high quality of information flows (see for example Lundvall 1993 and Powell 1990). In this respect it is argued that:

- If the performance of the partners in the co-operative arrangement is mutually rewarding and parties are interested in each others' success, this may motivate both open communication and the supply of relevant information to each other.

- Members of a co-operative relationship are better able to supply "good" and appropriate information to each other because they have better knowledge of the needs of their partner than is the case in spot-market relationships. This better knowledge of information requirements also enables them to filter the information relevant for their partner.

- As far as there is some reputation and trust involved, the quality of information received can be assessed much more easily than in spot-market relationships.

- Long term relationships may lead to some degree of inter-organisational adaptation with regard to the interfaces of the exchange partners. Therefore, information flows between the partners may be faster and less subject to errors than is the case in a spot-market relationship.

- If co-operation involves some trust or knowledge about the respective partner, this may result in a reduction of uncertainty with regard to the partner's future behaviour (Thorelli 1986; Galaskiewicz 1985).

All this leads to the hypothesis that relatively rich information flows within networks may not only accelerate a relatively early adoption of innovations and ideas but also stimulate the generation of new ideas concerning improvements of products and processes (cf. Saxenian 1994; Storper 1992).

2.5   Conclusions and Policy Implications

I have argued here that a basic characteristic of innovation networks is that many of its members have succeeded in overcoming obstacles of a division of innovative labour to a certain degree. In many cases, this requires the establishment of long-term relationships which enable the parties to deal with the problems associated with asymmetric information and incompletely specified contracts. Reviewing the various possible advantages of networking, it becomes obvious that most of such advantages result from the redundancy of vertical relationships and not from horizontal co-operation. This finding is a remarkable contrast to large parts of the literature on networks in which the main emphasis is on horizontal relationships. We may, therefore, conclude that the importance of horizontal relationships tends to be overestimated in this literature.

Stimulating contacts and trying to help economic actors find an appropriate partner for a division of labour would be one strategy to support the emergence and the development of innovative networks. Possible measures could be providing information about potential partners for R&D co-operation and creating opportunities for contact and decentralised exchange of information. Real world examples of effective networks reported in the literature suggest that the provision of a resource that is jointly used by the members of the network (e.g., a public research institution) may also play an important role in the formation of clusters of firms that later co-operate in networks. However, we still do not know much about promising ways to generate and stabilise networks or networking behaviour of the type under review here. Considerable further research will be necessary before this question can be answered satisfactorily.

2.6   References

CAMAGNI, R. (Ed.) (1991): *Innovation Networks: Spatial Perspectives*, London: Bellhaven.

CARR, J.L./LANDA, J.T. (1983): The Economics of Symbols, Clan Names, and Religion, *Journal of Legal Studies*, 12, pp. 35-156.

COHEN, W./LEVINTHAL, D.A. (1989): Innovation and learning: The two faces of R&D - implications for the analysis of R&D investment, *Economic Journal*, 99, pp. 569-596.

GALASKIEWICZ, J. (1985): Interorganisational Relations, *Annual Review of Sociology*, 8, pp. 281-304.

GRABHER, G. (Ed.) (1993): *The embedded firm – On the socioeconomics of industrial networks*. London: Routledge.

GRABHER, G. (1993): The weakness of strong ties: the lock-in of regional developments in the Ruhr area, GRABHER, G. (Ed.): *The embedded firm – On the socioeconomics of industrial networks*. London: Routledge, pp. 255-277.

GRANOVETTER, M. (1973): The Strength of Weak Ties, *American Journal of Sociology*, 78, pp. 1360-1380.

KATZ, M.L./ORDOVER, J.A. (1990): R&D Co-operation and Competition, *Brookings Papers on Economic Activity – Microeconomics*, pp. 137-203.

LUNDVALL, B.-Å. (1993): Explaining interfirm co-operation and innovation - Limits of the transaction-cost approach, GRABHER, G. (Ed.): *The embedded firm – On the socioeconomics of industrial networks*. London: Routledge, pp. 52-64.

MACNEIL, I.R. (1978): Contracts: Adjustment of Long-term Economic Relations under Classical, Neoclassical and Relational Contract Law, *Northwestern University Law Review*, 72, pp. 854-905.

PORTER, M. (1998): Clusters and the new economics of competition, *Harvard Business Review*, November-December, pp. 77-90.

POWELL, W.W. (1990): Neither Market Nor Hierarchy: Network Forms of Organisation, *Research in Organisational Behaviour*, 12, pp. 295-336.

PYKE, F./BECATTINI, G./SENGENBERGER, W. (Eds.) (1990): *Industrial districts and inter-firm co-operation in Italy*. Geneva: International Institute for Labour Studies.

SAXENIAN, A. (1994): *Regional Advantage*. Cambridge (MA): Harvard University Press.

STORPER, M. (1992): The limits of globalisation: technology districts and international trade, *Economic Geography*, 28, pp. 60-93.

THORELLI, H.B. (1986): Networks: Between Markets and Hierarchies, *Strategic Management Journal*, 7, pp. 37-51.

VON HIPPEL, E. (1987): Co-operation between Rivals: Informal Know how Trading, *Research Policy*, 16, pp. 291-302.

VON HIPPEL, E. (1994): "Sticky information" and the locus of problem solving: implications for innovations, *Management Science*, 40, pp. 429-439.

WEICK, K.E. (1976): Educational Organisations as Loosely Coupled Systems, *Administrative Science Quarterly*, 21, pp. 1-19.

Section II: **Knowledge and Learning in Innovation Networks**

(Germany)

3. Knowledge, Innovation Processes and Regions

Emmanuel Muller

031

032

D83

R11

3.1 Introduction

The expanding field of economic literature devoted to knowledge production and diffusion strongly emphasises that knowledge is increasingly becoming a crucial resource for growth. Moreover, the issues of knowledge and innovation appear as intimately inter-related, underlining their decisive influence for the competitiveness of firms and countries, but also for the development and prosperity of regions. In fact, and this is only a paradox at first glance, despite (and even to a certain extent, due to) its intangible nature knowledge is not ideas floating in a purely abstract vacuum but is rooted in the economic reality, and is thus, at least partially, linked to territories. This paper aims at highlighting the complex relations between knowledge, innovation and regions. In the first section, a brief theoretical overview provides the key conceptual elements which allow the mechanisms of knowledge creation and diffusion and their implications for innovation and regional development to be questioned. The second section of the paper offers an illustration of how knowledge exchanges between different categories of actors may take place. The proposed typology displays some stylised facts related notably to the spatial patterns of innovation interactions. Finally, the concluding part raises several issues which can be considered from the researcher's as well as from the policy-maker's points of view.

3.2 From Knowledge to Regions: a Theoretical Overview

This first section is devoted to the theoretical aspects of the interrelations between knowledge, innovation and regions. At first, the difference between knowledge and

The present paper has been written with support from the German Research Association (Deutsche Forschungsgemeinschaft, Programme "Technological Change and Regional Development in Europe") and from the EU programme TSER (TIPIK project: Technology and Infrastructures Policy in the knowledge-based economy - the Impact of the technology towards codification of Knowledge).

information is established, which is completed by the distinction between tacit and codified knowledge. This leads to a consideration of innovation, in line with the evolutionary approach of innovation, as a cycle associating tacit and codified knowledge within firms and between firms and further actors. Finally, the issues of knowledge sharing and innovation interactions on a regional level are addressed.

Dealing with knowledge implies a critical attempt to define its nature. In this respect, it seems particularly important to stress that knowledge is more than a sum of information. Nonaka (1994) underlines the necessity to distinguish clearly between "information" and "knowledge", although they are sometimes used as synonyms. In this respect, information can be considered as a flow of messages or meanings *which might add to*, *restructure* or *modify* knowledge. Information is thus a necessary and inseparable *medium* for establishing and formalising knowledge. More precisely, one may refer to the distinction provided by Laborit (1974), between: (i) circulating information, (i.e. routinised and repetitive information); and (ii) structuring information, (i.e. information liable to provoke or to favour an adaptation or a transformation and leading to non-routinised decisions). This distinction (cf. Ancori 1983) reviews in fact two visions of communication: the *Shannonian* approach (in which information is reduced to the status of signals, and communication consists of the transmission of these signals) and the *Batesonian* one. In this latest approach, informational flows which lead to "alterations" of the system are introduced in the analysis. Considering for instance a firm, its aptitude to organise its informational flows and to "extract them from/combine them with" its informational stock can be seen as the expression of its *knowledge-base*. This perspective allows the establishment of the distinction between information and knowledge. At the same time, it highlights the economic importance of knowledge, and consequently of learning. As Cooke (1998: 8) put it: "*Knowledge plays a fundamental role: the constitution of a firm is mediated by the knowledge possessed by the founder or "creative agent" and developed by learning. Firms learn from their own experience, but also from other firms they work with and with whom they share information, knowledge and technologies.*"

The economic understanding of knowledge relates primarily to the process of knowledge creation and diffusion within the economy. To grasp this process, it is necessary to consider the issues of codability and codification of knowledge. The distinction between tacit and codified knowledge has been established by Polanyi (1966). Whereas *codified knowledge* is easily transmittable in a formal and systematic language (comprising words, figures, etc.), *tacit knowledge* always has an implicit or individual related character (strongly based on personal experience) which makes its formalisation and exchange difficult. The "knowledge pyramid" provides a possible representation (among others) of the distinction between tacit and codified knowledge (cf. Figure 3-1). Nevertheless, it is important to keep in mind that these two forms complement rather than substitute each other. In fact, they tend to co-evolve: the process of codification generates new tacit knowledge. For instance, the ability of an individual to understand or interpret the codes in which knowledge is articulated is itself based on tacit knowledge, which can only be acquired through practice and experience. "*Any organization that dynamically deals with a changing environment ought not only to process information efficiently, but also create information and knowledge*", asserts Nonaka (1994: 14). This circulation of knowledge implies its transformation along two dimensions (explicit/tacit - individual/social). Focussing on firms and on innovation activities, the knowledge-base of a firm can be interpreted as a combination of *tacit* (or implicit) and of *codified* (or explicit) knowledge. The expansion of a firm's knowledge base can be realised by the exploitation of internal search capacities or by the acquisition of external knowledge (cf. Saviotti 1998). In this respect, a firm's expansion depends on the "absorptive capacities" it develops (cf. Cohen/Levinthal 1989). To sum up, knowledge constitutes a pre-condition for understanding (new) information; and to create (additional) information. Consequently, knowledge is intimately interrelated to innovation processes.

Figure 3-1:   **The pyramid of knowledge**

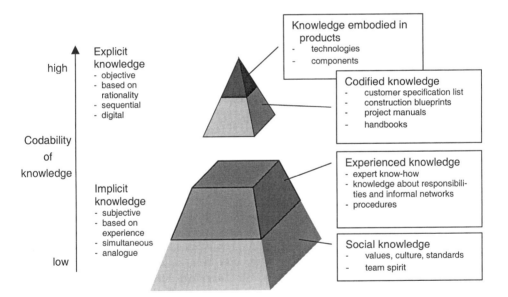

Adapted from: Gassmann (1997: 152)

The phenomenon of innovation can be conceived as an evolutionary process based on knowledge. This vision corresponds to the approach adopted by evolutionary economics (cf. Nelson/Winter 1974, 1975, 1977). The central role played by knowledge for innovation is ideally depicted by Kline and Rosenberg (1986) in their "*chain-linked model*". This model can be interpreted on different levels: while the (traditional) linear model features only one single level, the "chain-linked model" comprises five of them (cf. Figure 3-2).

Figure 3-2: **The chain-linked model**

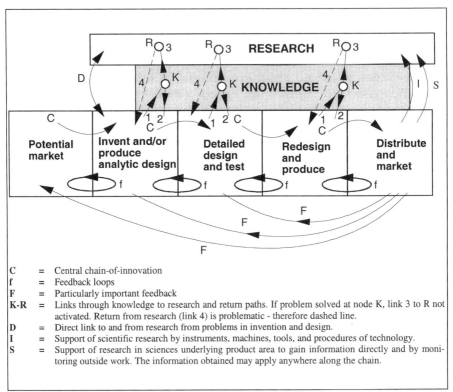

C	=	Central chain-of-innovation
f	=	Feedback loops
F	=	Particularly important feedback
K-R	=	Links through knowledge to research and return paths. If problem solved at node K, link 3 to R not activated. Return from research (link 4) is problematic - therefore dashed line.
D	=	Direct link to and from research from problems in invention and design.
I	=	Support of scientific research by instruments, machines, tools, and procedures of technology.
S	=	Support of research in sciences underlying product area to gain information directly and by monitoring outside work. The information obtained may apply anywhere along the chain.

Source: Kline/Rosenberg (1986: 290)

The first level consists of the central chain which - extending from the "conception" to the "distribution" of innovation - can be seen as the integration of the linear model. The second level is constituted of the feed-back loops associating phases of the central chain: each stage is linked to the preceding one, and the last (marketing and distribution) to all the other phases. The third level links the central chain of innovation to research[1] which constitutes a kind of "knowledge-stock" likely to stimulate each step of the innovation process (and not only its beginning, in contrast to the linear model). The last levels are less frequent and feature situations where respectively: (i) dramatic scientific shifts generate radical innovations; and (ii) innovations support the expansion of scientific knowledge.

---

[1] The term *research* encompasses internal and external activities.

The phenomenon of innovation should be understood as a *cycle involving interaction between tacit and codified knowledge*. To make the link with the approach adopted by evolutionary economics, it is possible to assert that: (i) firms are organisations which apply different inputs, one of the most relevant for innovation being *information*; (ii) information is *accumulated in* and *processed by* the knowledge base of the firm; and (iii) *knowledge accumulation* and *knowledge processing* by firms results from learning. From a dynamic perspective, knowledge can be seen as expanding by associating in different forms, tacit and codified knowledge.[2] On the one hand, the codification of tacit knowledge allows an availability of knowledge which increases with time. On the other hand, the dynamic expansion of codified knowledge generates the apparition of new areas of tacit knowledge (cf. Figure 3-3).

Figure 3-3:   A context of expanding knowledge

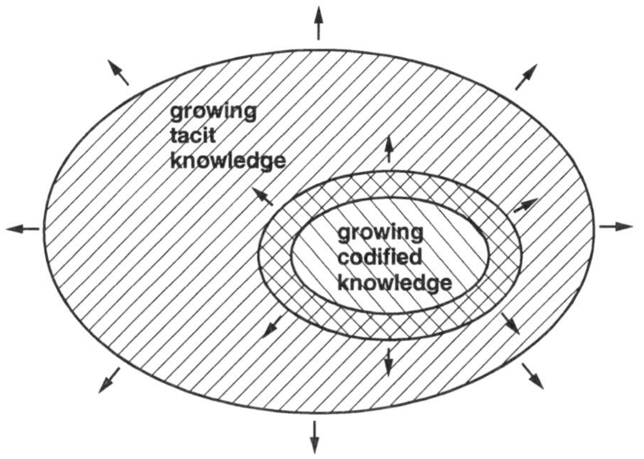

Source: Muller *et al.* (1998: 6)

Three further observations may be relevant before introducing the regional dimension in the analysis of the relations between knowledge and innovation. Firstly, the above developed conception of the knowledge base of the firm authorises (and furthermore implies) the existence of *shared learning effects*. Shared learning effects correspond to situations where learning occurs simultaneously within and outside a given organisation, for instance, when learning takes place in two firms at the same time. It can quite easily be assumed that shared learning effects may influence the innovativeness of firms. The second observation concerns *the growing possibilities*

---

2   For a detailed analysis of knowledge creation, transformation and diffusion within firms, see Nonaka (1994).

of using new means of codification in the production of knowledge, notably (but not exclusively) due to the dramatic development of information and communication technologies (ICT). These new means of codification affect notably: (i) the generation of new knowledge by firms; (ii) the storing of and access to top knowledge within firms; and (iii) the exchange between firms and further actors. Moreover, these three are strongly inter-related and mutually reinforcing. Finally, considering that the development of a firm's knowledge base (through learning) determines its competencies (and thus its innovativeness), it appears that the *learning process is path dependent*. In fact, the development of competencies by a firm is path dependent in the sense that: (i) a firm learning consists of the combination of "knowledge items" taking different forms (for instance tacit and codified) which implies having been previously acquired or generated ; (ii) this combination refers to an application, i.e. a further step in the process. This remark seems valid even if one considers particular cases in which, for instance, organisational forgetting is a necessary step of the learning process.

Regional learning directed towards innovation is not simply the acquisition of information by firms or groups of firms located within a specific territorial unit defined as a region. It may seem more judicious to consider regional learning as a process by which information available (inside and outside the region) to regional actors becomes usable knowledge for firms located in the region. In this respect, the *role and impact of innovation systems* can be examined by distinguishing mainly two levels: the national and the regional one. Systems may be defined in this context as "*complexes of elements or components, which mutually condition and constrain one another, so that the whole complex works together, with some reasonably clearly defined overall function*" (Edquist 1997: 13). In the continuation of Freeman (1987) and Lundvall (1988, 1992) abundant literature can be found related to national innovation systems (NIS). The regional innovation system (RIS) approach (cf. Cooke *et al.* 1996; Cooke 1998) encompasses the concepts of "industrial district", "innovative milieu" and "regional learning" to the greatest extent. It allows a summary of the arguments in favour of an influence of the regional environment on firms' innovation capacities.[3] RIS can be perceived as a transposition of NIS on the regional level. From both regional and national perspectives, the system of innovation is constituted by elements and relationships interacting in the production, diffusion and use of new knowledge. Thus, it provides a set of arguments highlighting why and how certain regions (and *a fortiori* some countries) are more innovative than others.

---

[3] An alternative perception, which can be found in the literature but is not retained here, should be briefly evoked. In fact, it is possible to define a region from an economic perspective, for instance with the help of the approach in terms of clusters (in the sense of industrial clusters given by Porter 1990). From this point of view, the industrial cluster may be seen as the sum of all the economic actors contributing directly to the dominant production process of the considered region.

In fact, considering innovative regions, the previously evoked circulation and combination of knowledge is achieved by two main (non–exclusive) methods: (i) the activity and mobility of skilled personnel; and (ii) the development of formal and informal relations of co-operation among actors. In both cases, the ability to transfer knowledge within the region (and to a certain extent the aptitude to catch up with knowledge from outside the region) will depend on a common language and on a shared knowledge base. In this respect, as Keeble/Wilkinson (1999: 299) put it: *"Tacit knowledge (...) is specific to organizational and geographic locations and this increases its internal circulation but impedes its external accessibility."* In other words, localised innovation networks favour an efficient circulation of codified knowledge since the structure of the interactions makes different segments of specific tacit forms of knowledge compatible. More generally, referring to the arguments developed by Lawson/Lorenz (1999), the processes underlying regional collective learning can be examined along three central ideas. Firstly, the overall attitude towards innovation within a region, which relies on the concept of *regional innovation culture*. A lack of innovation culture has as a consequence developed regional inertia, meaning that firms face resistance when trying to introduce, to use and to diffuse new knowledge.[4] Secondly, the resources potentially available within a region in terms of knowledge bases and possible knowledge combination constitute the *regional innovation potential*. As exposed previously, learning, notably with respect to innovation, is a path dependent and cumulative process. The search for and access to new knowledge which becomes incorporated into the regional firm's routines in an incremental manner enables the introduction of innovations (*i.e.* mostly incremental and exceptionally radical innovations). Finally, the willingness of regional actors to be open to knowledge sharing and their ability to exchange ideas on the issue of innovation constitute the aptitude to build up *regional innovation networks*. The existence of a "common language" or, more generally, of a business and regionally rooted culture explains that such networks may resolve the problems of co-ordination/competition which inevitably arise when knowledge becomes shared on innovation issues.

Considering the *policy implications of regional learning* in terms of innovation support, it seems clear nowadays that the nature and frequency of interactions between actors, notably in the form of networks, influence regional development (cf. for instance Braczyk *et al.* 1998). Nevertheless, and this probably constitutes one of the main policy challenges for European regions, no unique way to foster regional innovation capacities and to ensure some competitive advantage can be established. In this context, the multiplication of innovation-related initiatives, notably initiated by the European Union, can be interpreted as a willingness to promote the conditions for sustained knowledge-based economic development on a regional level (for

---

4 To a certain extent, it is possible here to make the link with the GREMI (*Groupe de Recherche Européen sur les Milieux Innovateurs*) approach related to "innovative milieux" (cf. for instance Perrin 1990; Maillat and Perrin 1992).

an introduction and overview of regional innovation-related initiatives supported by the European Union, cf. for instance Landabaso, 1997, and Larédo 1995). Learning is the key component of those policy tools. In fact, such initiatives may be considered as a form of social engineering, since they aim at the establishment and development of what Landabaso *et al.* (1999) call the "social capital" of a region. "*The more a region (or a company) is in a position to learn (identify, understand and exploit knowledge [...]) the more capable, and possibly willing, it becomes to build on and increase its demand and capacity to use further new knowledge.*" (Landabaso *et al.* 1999: 6). Considering the issue of regional competitive advantages, localised innovation networks (which are rooted in the region, even if those networks include actors from "outside", which can be desirable and useful) appear as even more decisive than the skills of the regional labour force (since highly skilled personnel may "escape" from the regional system). In fact, for a region, the establishment and development of a "social capital" is a way to master capabilities which cannot simply be transferred or easily replicated elsewhere. In this respect, the co-existence (within a country, a continent or even in a globalised world) of various types of regional environment, corresponding to different innovation-related infrastructural qualities, implies specific functional organisation of knowledge-related activities. Nevertheless, this does not necessarily imply that regional inequalities in terms of innovation infrastructure cannot be successfully compensated for, refuting the belief in "territorial fatality" (Muller 1999a).

3.3 An Illustration of the Influence of Spatial Patterns on Knowledge Interactions

A comprehensive theory-oriented literature can be found concerning the links between knowledge and innovation on a regional level; nevertheless these reflections are only supported by parcelled and heterogeneous empirical studies. This second section of the paper provides an illustration of knowledge interactions and of their spatial patterns. This example is extracted from a broader body of empirical investigation dealing with regional innovation potential and focusing notably on the examination of interlinkages between three types of actors: (i) manufacturing firms (and in particular SMEs); (ii) knowledge-intensive business services (KIBS); and (iii) universities and research institution labs (also called institutions of technical infrastructure or ITI).[5] The illustration below corresponds to a part of the survey

---

5 This operation entitled "Analysis of regional innovation potential from a European perspective" (1994-1998) was granted by the *Deutsche Forschungsgemeinschaft* (DFG, the German Research Council) and designed conjointly by four research teams. The joint research project involved the Department of Economic Geography at the University of Hanover, the Chair for Economic Policy at Technical University Bergakademie Freiberg, the Department of Economic and Social Geography at University of Cologne, and the Fraunhofer Institute for Systems and Innovation Research (FhG-ISI) in Karlsruhe. A recent special issue of *European Planning Studies* (Vol. 8, No. 4, 2000) presents different aspects of the methodology and of the results of the overall project.

performed for the border regions of Alsace and Baden. In terms of forms of knowledge involved, this area is particularly interesting for the analysis, since it corresponds to territories which belong to two different national innovation systems, and contains five (administrative) sub-units presenting different resources in terms of innovation–related infrastructures (cf. Figure 3-4). The aim here is not to display detailed results specific to this area, but to try to benefit from the empirical results accumulated in order to go one step further in the reflection about knowledge, innovation processes and regions.[6] More precisely, the consequences in terms of innovation and the spatial patterns of knowledge interactions between different categories of actors are considered.

Figure 3-4:   **The surveyed regions**

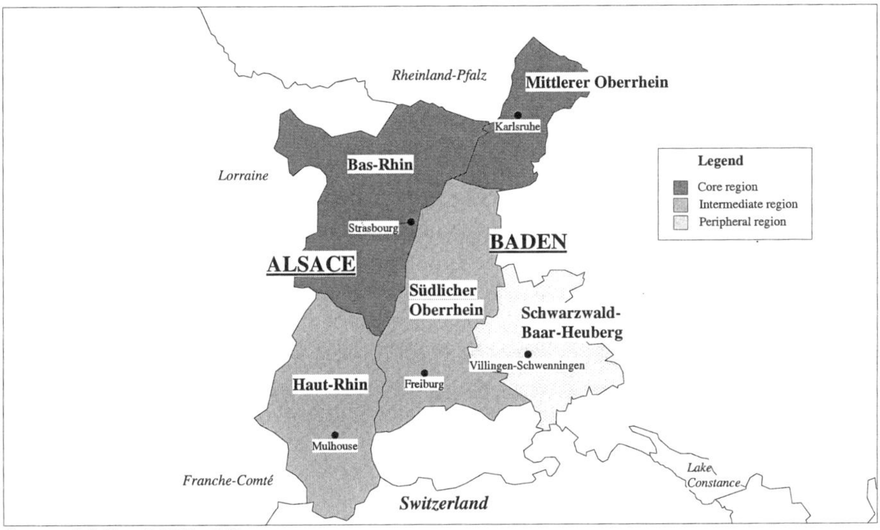

Source: Muller (1999b: 85)

The typology of innovation interactions proposed hereby as an illustration is entitled "the wheel of knowledge interactions" (cf. Figure 3-5). It aims at featuring knowledge exchanges encompassing:

(i)   manufacturing SMEs;

(ii)   KIBS;

(iii)   institutions of technological infrastructure (ITI);

(iv)   large manufacturers (*i.e.* not small and medium-sized firms); and

(v)   further service firms (*i.e.* non-KIBS).

---

6   Detailed empirical results dealing with the case of Alsace and Baden can be found in Koschatzky (1997), Muller (1997), Muller/Zenker (1998), Héraud/Muller (1998), Muller (1999b), Koschatzky (1999, 2000).

A set of approximately 40 personal interviews performed in Alsace and Baden contributed to the establishment of this typology (cf. Muller 1999b: 154-159).[7] The issues of the role and impact of KIBS and ITI are the object of further attention in the present literature: see notably Strambach concerning KIBS and Héraud/Bureth (already referred to in this volume) for analysis related to ITI. The "wheel of knowledge" constitutes a basic representation displaying in a speculative manner some stylised facts about knowledge exchanges. Seven types of interactions or "links" are schematically depicted according to (i) the type of knowledge involved; (ii) the spatial patterns of the considered interactions; and (iii) the influence in terms of firms' innovations.

Figure 3-5:   **The wheel of knowledge interactions involving KIBS and SMEs**

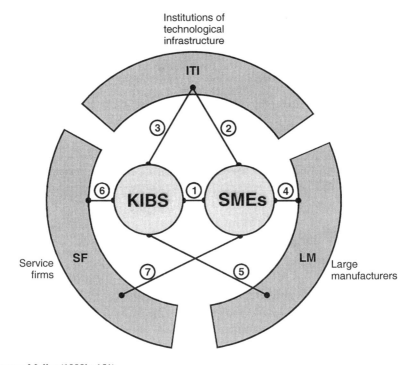

Source: Muller (1999b: 151)

---

7   The interviews, performed during the first semester of 1997, dealt with SMEs, KIBS, research labs and regional innovation-support organisations identified mainly on the basis of a postal survey. The aim was to gain additional qualitative information supplying the mainly quantitative character of the overall investigation (cf. Muller 1999b: 77-138).

(1) **Links between SMEs and KIBS**. This relation is characterised by the mutual impact of interaction between SMEs and KIBS on their respective innovation capacities. Considering empirical results (cf. Muller 1999b: 139-147), it can be assumed that: (i) when knowledge flows from SMEs to KIBS the knowledge transmitted is mainly of a tacit nature and proximity plays an important role; and (ii) when knowledge flows from KIBS to SMEs, proximity appears to be less important because the transmitted knowledge is mainly codified.

(2) **Links between SMEs and ITI**. The direct links between SMEs and ITI have as their main function to support and reinforce SMEs' innovation potential. It can be assumed that knowledge exchanges are most often of a codified nature. In this context, proximity between SMEs and ITI is probably not neutral (particularly due to knowledge transmission-related costs), but is nevertheless also not determinant: the specificity of the knowledge required is in fact the main selection factor.

(3) **Links between KIBS and ITI**. In general, the same basic assumptions can be advanced as for the relations, which are more common, between SMEs and ITI. One may even suppose that the need for (highly specialised) codified knowledge determines the links between KIBS and ITI and that, as a consequence, proximity only has little or no impact on those knowledge exchanges.

(4) **Links between SMEs and large manufacturers**. It is possible to speculate on the nature of the knowledge exchanges between SMEs and large manufacturers by comparing two cases schematically. The first case relates to innovations corresponding to the adoption by an SME of new techniques developed by large manufacturers. In this situation, the knowledge transmitted (for instance embodied in equipment) can be seen rather as codified than tacit and proximity as exerting little or no influence on the relation. The second case relates to specific development (of products, of processes, ...) performed by an SME in order to satisfy the requirements of a large manufacturing client. This corresponds for instance typically to sub-contracting situations. In comparison to the first case considered, it can be assumed that tacit knowledge potentially plays a greater role in the transmission of knowledge (from the SME to the large manufacturer) and that proximity favours the exchange of knowledge to a certain extent.

(5) **Links between KIBS and large manufacturers**. Three typical situations can be used to depict these relationships. The first situation is the adoption by a KIBS of artefacts produced by large manufacturers. The adoption induces associated organisational change (such as, for instance, related equipment in the case of information technologies). In such a situation, knowledge exchanges can be seen as mainly codified (strongly embodied in the artefacts) and relatively insensible to proximity effects. The second situation relates to knowledge exchanges taking the form of support (for instance managerial consultancy) provided by KIBS to large manufacturers. Such support constitutes a

potential source of (internal) innovation for KIBS. It can be interpreted as the application of knowledge which is partly tacit and partly codified. In this case, it is also realistic to consider that it only plays a marginal role. The third situation depicts KIBS resulting from the outsourcing of a specific activity initiated by a large manufacturer. In comparison to the two previous forms of relations, this type of interaction is characterised rather by the circulation of specific tacit knowledge and by the importance given to proximity between the partners.

(6) **Links between KIBS and service firms**. Service firms represent one of the most important groups of customers for KIBS. The effects of relations with (non-KIBS) service firms on KIBS' innovation capacities rely mainly on the concept of new services or on the evolution of existing ones (in order to fulfil service firms' emerging or changing needs). These relations can mainly be considered as a process of knowledge codification: (i) which takes place in the client firms; (ii) which is based on KIBS accumulated tacit knowledge; and (iii) for which proximity is rather unimportant.

(7) **Links between SMEs and service firms**. In contrast to the interactions involving KIBS, the knowledge exchanges which take place between SMEs and service firms correspond rather to routine services and thus have a relatively small impact on SMEs' innovation capacities. The type of knowledge involved and the role of proximity depends strongly on: (i) the degree of standardisation of the activity associating SMEs and service firms; and (ii) the importance of elements like trust for the service relation. Schematically, highly standardised services (typically based on the application of codified knowledge and by which the question of proximity is only related to cost effects) can be compared to more specific services (which imply a greater proportion of tacit knowledge and for which proximity is at least potentially important due to the necessity of a trusting relationship between the SME and the service firm).

3.4   Conclusion: Research and Policy Agenda

When trying to answer the question "what shall we do in the future to better understand the relations between knowledge, innovation processes and regions?", it seems important to proceed along three main axes: (i) the concepts on which forthcoming analyses will be based; (ii) the tools which may be useful for performing these analyses and (iii) the policies required for supporting knowledge development, notably on a regional level. New concepts are needed to respond to the challenging and complex issues raised by the drastic expansion of the place of knowledge in economic activities. It can reasonably be considered that only a reinforced movement of interdisciplinary reflections (associating notably epistemology, sociology and psychology to economics) may allow new concepts to flourish. Nevertheless, new concepts need to be accompanied by new tools, destined notably to improve measuring in a satisfactory manner, and thus to better understand the multiplicity of phenomena related to knowledge creation and diffusion. In this respect, investigation instruments such as multifactor analysis, neuronal networks and fuzzy indicators seem worth examining. Finally, new challenges call for new policies: a knowledge-ruled economy offers new perspectives for innovation-based regional development. In the European context especially, due to the double pressure of a reinforced need for regional convergence within the European Union and of the emergence of requirements linked to the possible expansion of the Union eastwords and southwords, regional development policies appear more decisive than ever for the future.

3.5   References

ANCORI, B. (1983): Communication, information et pouvoir, LICHNEROWICZ, A./PERROUX, F./GADOFFRE, G. (Eds.): *Information et communication*. Maloine:Verlag, pp. 59-84.

BRACZYK, H.J./COOKE, P./HEIDENREICH, M. (1998) (Eds.): *Regional Innovation Systems - The Role of Governance in a Globalized World*. London: UCL Press.

COHEN, W./LEVINTHAL, D. (1989): Innovation and Learning, the Two Faces of R&D, *Economic Journal*. 99, pp. 569-596.

COOKE, P. (1998): Origins of the concept, BRACZYK, H.J./COOKE, P./HEIDENREICH, M. (1998) (Eds.): *Regional Innovation Systems - The Role of Governance in a Globalized World*. London: UCL Press, pp. 2-25.

COOKE, P./BOEKHOLT, P./SCHALL, N./SCHIENSTOCK, G. (1996): *Regional Innovation Systems: Concepts, Analysis and Typology*. Paper presented during the EU-RESTPOR Conference "Global Comparison of Regional RTD and Innovation Strategies for Development and Cohesion". Brussels, 19-21 September.

EDQUIST, C. (Ed.) (1997): *Systems of Innovation. Technologies, Institutions and Organizations*. London: Pinter Publishers.

FREEMAN, C. (1987): *Technology Policy and Economic Performance: Lessons from Japan*. London: Pinter Publishers.

GASSMANN, O. (1997): *Internationales F&E-Management*. München: Oldenbourg Verlag.

HÉRAUD, J.-A./MULLER, E. (1998): *The Impact of Universities and Research Institutions Labs on the Creation and Diffusion of Innovation-Relevant Knowledge: the Case of the Upper-Rhine Valley*. Paper presented at the 38th Congress of the European Regional Science Association, August 28-31 1998, Vienna.

KEEBLE, D./WILKINSON, F. (1999): Collective learning and knowledge development in the evolution of regional clusters of high technology SMEs in Europe, *Regional Studies*. 33, pp. 295-303.

KLINE, S./ROSENBERG, N. (1986): An Overview of Innovation, LANDAU, R./ROSENBERG, N. (Eds.) (1986) *The Positive Sum Strategy: Harnessing Technology for Economic Growth*. Washington D.C.: National Academy Press, pp. 275-305.

KOSCHATZKY, K. (1997): *Entwicklungs- und Innovationspotentiale der Industrie in Baden*. Erste Ergebnisse einer Unternehmensbefragung. Karlsruhe: Fraunhofer-ISI (Arbeitspapier Regionalforschung Nr. 5).

KOSCHATZKY, K. (1999): Innovation Networks of Industry and Business-Related Services – Relations Between Innovation Intensity of Firms and Regional Inter-Firm Cooperation, *European Planning Studies*, 7, pp. 737-757.

KOSCHATZKY, K. (2000): A River is a River - Cross-Border Networking Between Baden and Alsace, *European Planning Studies*, 8, pp. 429-449.

LABORIT, H. (1974): *La nouvelle grille*. Paris: Laffont.

LANDABASO, M. (1997): The promotion of innovation in regional policy: proposals for a regional innovation strategy, *Entrepreneurship & Regional Development*, 9, pp. 1-24.

LANDABASO, M./OUGHTON, C./MORGAN, K. (1999): *Learning regions in Europe: theory, policy and practice through the RIS experience*. Paper presented at the 3rd international conference on technology and innovation policy: assessment, commercialisation and application of science and technology and the management of knowledge. Austin, USA: August 30 - September 2, 1999.

LARÉDO, P. (1995): *A preliminary characterization of RITTS*. Paper presented at the workshop Regional Innovation and Technology Transfer Strategies and Infrastructures. Luxembourg: May 31, 1995.

LAWSON, C./LORENZ, E. (1999): Collective learning, tacit knowledge and regional innovation capacities, *Regional Studies*, 33, pp. 305-317.

LUNDVALL, B.-Å. (1988): Innovation as an Interactive Process: From User-Producer Interaction to the National System of Innovation. DOSI, G./FREEMAN, C./NELSON, R./SILVERBERG, G./SOETE, L. (Eds.) (1988) *Technical Change and Economic Theory*. London: Pinter Publishers, pp. 349-369.

LUNDVALL, B.-Å. (Ed.) (1992): *National System of Innovation. Towards a Theory of Innovation and Interactive Learning*. London: Pinter Publishers.

MAILLAT, D./PERRIN, J.-C. (Eds.) (1992): *Entreprises innovatrices et développement territorial*. Neuchâtel: EDES.

MULLER, E. (1997): *Entwicklungs- und Innovationspotentiale der Industrie im Elsass*. Karlsruhe: Fraunhofer-ISI (Arbeitspapier Regionalforschung Nr. 8).

MULLER, E. (1999a): *There is No Territorial Fatality! (or How Innovation Interactions between KIBS and SMEs May Modify the Development Patterns of Peripheral Regions).* Paper presented at the 39th ERSA (European Regional Science Association) Congress. Dublin: August 23-27, 1999.

MULLER, E. (1999b): *Innovation Interactions Between Knowledge-Intensive Business Services and Small and Medium-sized Enterprises – Analysis in Terms of Evolution, Knowledge and Territories.* Ph.D. dissertation, Faculté des Sciences Économiques et de Gestion. Strasbourg: Université Louis Pasteur.

MULLER, E./ZENKER, A. (1998): *Analysis of Innovation-oriented Networking Between R&D Intensive Small Firms and Knowledge Intensive Business Services - Empirical Evidence from France and Germany,* Proceedings of the High-Technology Small Firm Conference. Twente: University of Twente, pp. 175-203.

MULLER, E./ZENKER, A./MEYER-KRAHMER, F. (1998): *The Consequences of a Growing Codification of Knowledge on the Evolution Capacities of European Firms and Regions.* Karlsruhe: Working document FhG-ISI.

NELSON, R./WINTER, S. (1974): Neoclassical vs Evolutionary Theories of Economic Growth. Critique and Prospectus, *Economic Journal,* December, pp. 886-905.

NELSON, R./WINTER, S. (1975): Growth Theory from an Evolutionary Perspective: The Differential Productivity Puzzle, *The American Economic Review,* 65, pp. 338-344.

NELSON, R./WINTER, S. (1977): In Search of a Useful Theory of Innovation, *Research Policy,* 6, pp. 36-76.

NONAKA, I. (1994): A Dynamic Theory of Organizational Knowledge Creation, *Organization Science,* 5, pp. 14-37.

PERRIN, J.-C. (1990): Organisation industrielle: La composante territoriale, *Revue d'Economie Industrielle,* 51, pp. 276-303.

POLANYI, M. (1966): *The Tacit Dimension.* New York: Doubleday.

PORTER, M. (1990): *The Competitive Advantage of Nations.* London: Macmillan.

SAVIOTTI, P.P. (1998): On the dynamics of appropriability, of tacit and of codified knowledge, *Research Policy,* 26, pp. 843-856.

4. Innovation Processes and the Role of Knowledge-Intensive Business Services (KIBS)

Simone Strambach

4.1 Introduction

The knowledge-intensive service industry is one of the most dynamic components of the service sector in Europe and in most highly industrialised countries. This dynamic growth can be seen no longer as a simple outsourcing phenomenon. It is an reflection of deep changes in production and organisational structures and it shows the increasing linkages and networks between economic activities.

Looking at the role of knowledge-intensive business services (KIBS) in innovation is a relatively new feature. In the last decade KIBS were not seen as a relevant group of actors in innovation research and they were not seen as an objective for innovation policy. A major reason for the lag in knowledge about the importance of KIBS for innovation is the present state of the statistics for services and innovation, which makes it impossible to establish the quantitative contribution these services make to innovation at both the macro and micro levels.

The paper focuses on the contribution of KIBS to innovation from a qualitative perspective, using the systems of innovation approach. The main argument is that KIBS play an increasing role in the production and diffusion of knowledge in the current globalising learning economy.

4.2 What Are Knowledge-Intensive Services (KIBS)? – Some Remarks about the Definition

In general terms, business services are those services demanded by firms and public institutions and are not produced for private consumption. KIBS are the most knowledge intensive of the business related services. Thus, they do not include such services as cleaning and maintenance and repairs. These "routine services" presumably play no role in stimulating innovation or producing qualitative spillover effects in the areas where they are provided.

Despite every effort, uniform definitions of firms or activities that can be classified as providing knowledge intensive business services are not available at the European level. Definitions are time related and the enormous structural changes that are taking place in this area of economic activities are partly responsible for the lack of transparency on the supply side.

Figure 4-1: **Common characteristics of knowledge intensive business services**

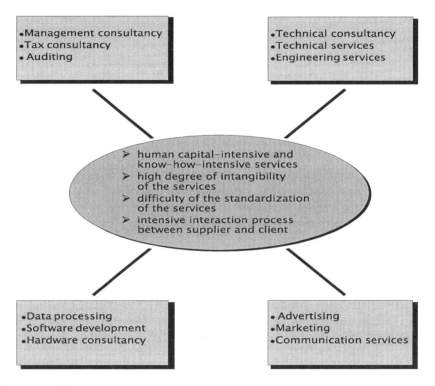

Source: own figure

The institutional conditions make rapid market entry possible in wide sections of this segment. There are now many firms supplying kinds of services which a few years ago did not exist. The lack of formal entry barriers in many branches of knowledge intensive services makes it possible to react quickly to current demand. However, firms which are new to the market are often unable to breach the informal barriers in the form of high problem solving competence and flexibility requirements. High entrance rates are connected with high living rates and thus structural dynamism is at a high level. The common characteristics of KIBS need to be emphasised, given the heterogeneity and the high structural dynamics of this service segment (cf. Figure 4-1).

There are three aspects which provide the links between KIBS. One is that the product of the firms is knowledge. The firms for the most part provide non-material intangible services. The use of the term knowledge intensive, analogous with the terms capital intensive and labour intensive, emphasises the fact that, for these firms, knowledge is the most important of all the factors of production. However, while capital and labour can be expressed in measurable economic units, the knowledge factor is difficult to grasp and even more difficult to measure. It must be stressed that knowledge is far more than just information. Starbuck (1997) states that knowledge is a stock of expertise and not a flow of information. Only the goal oriented integration of information represents knowledge. Unlike pure information, knowledge contains judgements, interpretations, and experiences and it is context dependent. Knowledge takes on different values in different situations and is not in any simple sense "objective". The fact that distinctions can be made between different kinds of knowledge is also important. Polanyi (1966) was one of the first to draw attention to the implicit and tacit dimensions of knowledge which are very important in the current discussion of knowledge management and the theoretical debates about organisational learning. Information and explicit knowledge can be systematically processed, transferred, and stored by organising it, implicit knowledge cannot. The latter is difficult to formalise, communicate, and transfer because it is either embedded in the culture of the organisation, in network relationships, or held only by particular individuals. These special features of knowledge make the standardisation of the non-material products of KIBS very difficult.

The second aspect that KIBS have in common is the very intensive interaction and communication which takes place between KIBS suppliers and KIBS users which is necessary for producing the services. The purchase of knowledge intensive services is not the same as the purchase of a standardised product or service. The exchange of knowledge products is associated with uncertainties and with information asymmetries in the quality evaluation stemming from the special features of the factor/commodity "knowledge".

The third important common aspect of all KIBS branches is that the activity of consulting, understood as a process of problem solving in which the KIBS adapt their expertise and expert knowledge to the needs of the client, makes up, in different degrees, the content of the interaction process between KIBS and their customers.

4.3 Changing Perspective on Innovation and the Firm – Towards a Systemic Understanding of Innovation

More recent findings of innovation research have greatly changed the understanding of technological change and they have also thrown a different light on the role and function of KIBS. Technological innovations and research carried out in formal structures still are important and very relevant where the technologies have a strong scientific basis, but this type covers only some of the complex innovation taking place at the end of the 1990s.

The acceleration of change in the globalisation process through market liberalisation and deregulation, through more complex and expensive technologies, and through the increasing diffusion of information, communication and knowledge, requires the enterprises to be highly flexible and adaptable. Firms must approach innovation in very broad sense, including technological development, marketing strategies and new work practices. The corporate capacity for continuous change must be increased dramatically. Managing innovation means for corporations the integration of technological, market and organisational innovations (cf. Tidd *et al.* 1997). Incremental innovations and organisational learning processes are of growing importance for the competitiveness of firms.

Studies made in the 1990s of national and regional innovation systems have shown that the environment in which innovation takes place is very important for the firms' innovation processes. These results have made a major contribution to shaping a systemic understanding of innovation processes (cf. Edquist 1997; Lundvall 1992; Nelson 1993; Braczyk *et al.* 1998.). The introduction of innovations not only depends on the competence and decisions of firms, it also depends on how knowledge originates and how innovations are adapted by actors situated in structures specific to particular countries and regions. Interactions between firms, and between firms and their environments, and the learning processes resulting from these interactions, are among the factors which explain a country's and a region's ability to innovate.

This broader understanding of innovation is very important in considering the contribution of KIBS to innovation and innovation processes.

4.4 Spatial Dimension: Growth Features of KIBS in Europe – National, Regional, Functional, and Sectoral Diversity

Numerous empirical studies of the spatial patterns of KIBS in different European countries show not only the dynamic growth but also that these have country specific features and that there are differences in KIBS segments from country to country (cf. Bade 1990; Illeris 1989; Moulaert/Tödtling 1995; Strambach 1995; Wood 1993). For example technical services are more important in Germany than in other countries.

Figure 4-2: **Share of business service employment in total employment in Europe 1996**

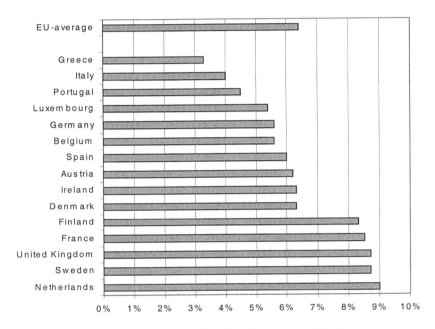

Source: Own calculation based on Labour Force data (Eurostat unpublished data)

Large interregional disparities show up clearly in the distribution of KIBS in European Countries (cf. Figure 4-2). On common trend is the concentration of KIBS in core metropolitan regions. In France, for example, the highest concentrations of KIBS is in the Ile-de-France and Rhônes-Alpes regions, in Great Britain in the south east, the region around London, and in Spain in the regions around Madrid and Barcelona. In Germany, too, the supply side is concentrated in regions with big agglomeration centres (see Figure 4-3). The fact that there are several of these concentrations reflects Germany's federal structure.

Investigations into the spatial organisation of KIBS show not only considerable regional disparities but also the different regional importance of individual branches in different agglomeration areas. A specific profile of the composition and pattern of the different KIBS branches can be seen in each of the individual regions. It seems obvious that region specific profiles and development paths do exist. The results of these analyses underline the importance of the influence of the institutional context.[1] The link with the recognition of national innovation systems comes in here. The development and specialisation of these services reflect the interaction between supply and demand which is shaped within the national and regional frameworks.

Research into both national and regional innovation systems has so far concentrated on the production system and on the emergence and diffusion of technological innovations and technological knowledge. The analyses of innovation systems mostly focus directly on institutions, organisations, and regulations connected with R&D. Suppliers of services in general, and of KIBS in particular, are not among the relevant groups of actors included in this research. The dynamic growth of knowledge intensive services in the present situation of structural change, however, appears to indicate that this segment is contributing to the production and diffusion of knowledge.

---

[1] Noyelle (1996: 23) argues, that the development of professional services is not directly linked to the country's level of development as measures by its income. Other elements, particularly national regulatory and historical factors, may explain national differences.

Figure 4-3: **Share of business services in total service employment in European regions 1996**

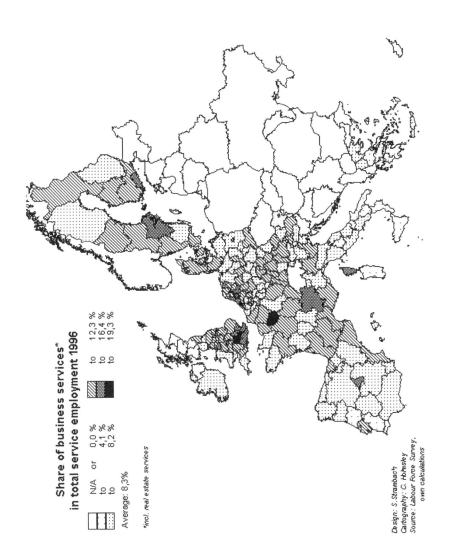

Source: Own map based on Labour Force data (Eurostat unpublished data)

4.5   The Contribution of KIBS to Innovation within National and Regional Innovation Systems: Direct and Indirect Effects

Analytically, the contribution KIBS makes to the innovation system can be divided into two parts which are very interdependent. On the one hand, there are the direct effects for competitiveness of national and regional economies, resulting from the innovative activities of the KIBS firms. On the other hand, KIBS also have indirect effects and positive feedbacks on the demand side which can emerge through the use by clients of the services of KIBS firms. Successful transfer of knowledge to clients, or innovative problem solving for them, can increase their competitiveness (cf. Figure 4-4).

Figure 4-4:   Contribution of KIBS firms in innovation systems

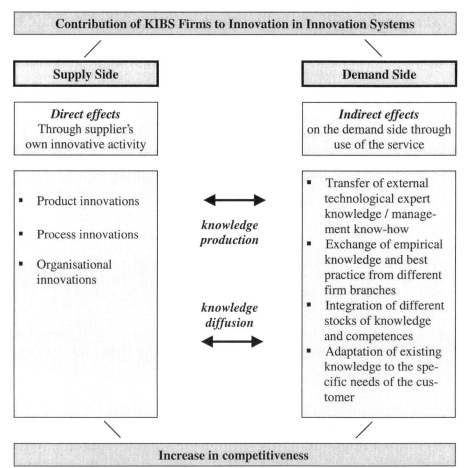

Source: Own figure.

4.5.1 Direct Effects – Innovation Activities of KIBS Firms

Innovation activities of KIBS are related to process, product or organisational innovations as well as in manufacturing industries. The increasing tradability of knowledge intensive services in the last decade and the technological possibility of transmitting knowledge intensive services over long distances foster the internationalisation not only of large KIBS firms but also of KIBS traditionally directed towards national and regional markets (cf. O'Farrell *et al.* 1998). Thus, in the globalisation process, knowledge products like management innovations or computer based consultancy products are also becoming important for the value creation and competitiveness of national and regional economies.[2] Empirical surveys made it evident that the innovative contributions of the different service branches need to be looked at individually. The EDP branch is marked by a high degree of product innovations whose implementation by the client can again effect process innovations. The software branches and the technical consulting are not only users of technologies, they also have a key function in the transfer and diffusion of technological innovations.

Making quantitative statements about the productivity and innovativeness of KIBS is difficult. Indicators and traditional tools used to evaluate productivity and innovativeness in the production system can only be used to a very limited extent for services. The reasons for this are the following:

- The traditional R&D concept has been shaped by technological innovations in the manufacturing sector.

- The internal innovation and knowledge organisation is as a rule only weakly formalised in service firms.

- In contrast to manufacturing firms, most KIBS firms do not distinguish R&D activities in organisational terms: quantitative indicators of innovation inputs like investment and employment in the R&D area therefore cannot be used.

- Patent applications are of limited use as output indicators for service firms. The reason is the extremely short innovation cycle of KIBS products.[3] In addition, the KIBS firms' advances in know–how are difficult to protect by patents, as they are largely personnel and context bound.

- It has not yet been possible to show investment in intangible capital assets in the statistics.[4]

---

2 An example from Baden-Württemberg is the firm SAP which became the market leader in the international markets, in a short time with its innovative product, a software solution called R/3:

3 At present the innovation cycle is only six months in some areas of the software branch.

4 For details of the problem of measuring non R&D innovation expenditures in surveys see Brouwer/Kleinknecht 1997: 1236

Empirical results for service innovations at the international level, covering the whole range of market oriented services, clearly underline the fact that service innovations can only be compared with those in the manufacturing sector in a very limited sense. There are differences both in the input factors of innovations and in the innovation processes themselves (cf. Brouwer/Kleinknecht 1997; Licht *et al.* 1997; DIW 1998). It is also evident that, because of the heterogeneity of the service branches, standardised surveys do not do justice to the complexity and the specific characteristics of the innovation processes of knowledge intensive business service firms.

4.5.2   Indirect Effects – Knowledge Diffusion and Knowledge Production in the Interaction Process

The strategic significance of KIBS in the innovation systems stems primarily from indirect effects and positive feedbacks that in the long run can increase the ability of the demand side to adjust and thus contribute to improving competitiveness.

These effects are the results of successful interaction and learning processes between KIBS suppliers and KIBS users, shaped and influenced by the institutional context. The role of KIBS in the national and regional innovations is closely tied to the "products" these services supply to the market. Specialised expert knowledge, research and development ability, and problem solving know–how are the real products of KIBS.

In innovation systems KIBS take on an important function in the *transfer of knowledge* in the form of expert technological knowledge and management know-how. As a result of the increasing differentiation and acceleration of the growth of knowledge and information and the vertical disintegration of the firm's functions, the more complex co-ordination of the changes in the firm's functions, production and sales requires specialised know-how not only for technological innovations but also for organisational changes. The growing national and international division of labour, the increase in knowledge, and the resulting concentration on the core competences of firms means that the competence of any individual firm is becoming narrower. Because the half life knowledge is becoming shorter, it will be increasingly difficult for firms to provide abundant intra organisational know-how in all the relevant areas and to keep up with the latest developments.

On the other hand, independent KIBS firms, which must establish themselves and ensure their survival with new products in the national and international markets, are faced with innovation, time and quality competition which forces them to continually build up new competences in their fields of knowledge. It should be stressed that, as a result of this development, the integration of external and in-house knowledge and the integration and use of competences is becoming much more important for innovative changes and problem solving. KIBS firms, as sources of external knowledge, have, with their skills and their competences, become involved to a far larger extent than before in the modernisation and rationalisation of production, management and sales of firms.[5] They also contribute to the diffusion of knowledge by spreading the knowledge gathered from experience and "best practice" to firms in different branches.

The *integration of different stocks of knowledge* and competences is a major function which KIBS fulfil in innovation systems (cf. Figure 4-5). Innovations and innovative problem situations in firms normally affect more than one functional area and thus the integration of the service product in the organisation of the customer firm of KIBS requires a high level of complementary knowledge as well as their own special core competences. The transfer of organisational innovations, like the process oriented reorganisation in particular, requires not only the technological knowledge to analyse and evaluate the firm specific structures and processes, it also requires management knowledge and social competences for following the process of change and for successful implementation.

The function of knowledge integration of KIBS is also reflected in the innovative products of these firms which can less and less often be assigned to a single branch. The integration and combination of different fields of knowledge, which were previously provided by separate branches, is a characteristic of innovative KIBS products and services. The combination of separate different disciplinary fields of knowledge is necessary to create innovative products and services.[6] The fields of engineering knowledge and EDP and IT know–how overlap. In addition, the border between traditional fields of management consultancy and software design are increasingly blurred.

Empirical studies in European countries show that formal and informal networks and co-operations have a key function for KIBS. That is another indicator for their integration function. Through these forms of working together they are able to cope with the conflict involved in having to be both specialist, and generalist, to unite depth and breadth of knowledge, and to unite separate disciplinary fields of knowledge in the process of solving problems for their customers.

---

5 See for example the results from Wood (1996) for the United Kingdom.

6 This is shown for example by an empirical study of technical services in the innovation system of Baden-Württemberg (Strambach 1999).

Figure 4-5:   **Knowledge production and knowledge diffusion in the interaction process of business services and their clients**

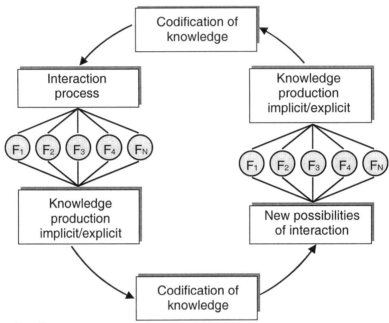

Source: Own figure

A further role played the KIBS in the innovation process is the *adaptation of existing knowledge* to the specific needs of the customer. Cohen/Levinthal (1990) have
shown that the successful use of external knowledge is influenced by the firms' existing knowledge structure. The changed competitive and market conditions have
meant that more technological combination opportunities are available than in previous years and that organisational innovations cannot be implemented as all purpose models or as closed concepts. Successful transfer of knowledge is tied up with
learning processes and the acquisition of new knowledge is smoother and less
problematic when the innovative knowledge can be added to the knowledge structure already available within the organisation. The fact that the relationships KIBS
have with their customers tend to be long term indicates their role in adapting the
existing knowledge to the firm specific situation. They acquire explicit and tacit
knowledge about the customer firm which makes it possible for them to adapt innovative problem solutions to the organisation specific requirements and to integrate
them into the corresponding firm structure and culture (cf. Strambach 1995).

In innovative systems KIBS not only perform the functions of transferring, integrating, and adapting knowledge, they also *produce new knowledge*. Unlike material goods which are used up during consumption, consuming knowledge actually
increases it. KIBS receive knowledge in the course of the interaction process that

takes place when the service is provided. This knowledge, which is created in the context of use, is not available with scientific institutions and is to a large extent tacit knowledge. It is collected, rearranged, and organised by KIBS by turning it into new products. In turn, these new knowledge products open up new opportunities for KIBS to interact with their customers. In a certain sense, KIBS create their own markets. Two processes are important here for achieving economies of scale and scope – the codification of knowledge and the ensuing standardisation of consulting procedures and consulting products.

4.5.3 The Rise of Different Roles for Large and Small KIBS in Innovation Systems

Although the interaction processes are shaped within national and regional framework conditions, some typical features of the relationships involved in the interactions between the KIBS and their clients become visible. The KIBS segment is highly segmented in the European countries with a few large, mainly multinational, KIBS firms and a large number of national and regional based small and medium-sized ones.[7] The big multinational KIBS, which mainly work for big internationally operating companies and have more recently also tapped the area of the large medium-sized firms, are now developing into what can be called a knowledge industry. Growing competitive pressure from internationalisation of both the customer and the service markets have led to extremely large concentration and expansion trends for multinational KIBS. The firm size will become more and more important for survival in global markets.

The importance of the large multinational firms in innovation systems stems primarily from the fact that they develop new consulting products in the form of methods, instruments, and models based their own know–how and experience. Unlike the smaller national or local suppliers, in many cases they have now formally set up internal R&D functions which further the creation of new expertise and codification processes of tacit knowledge. Transforming consulting product innovations into standard products occurs more quickly when it is carried out within an formal organisation. In this way, the large international firms hasten *the standardisation process* in the areas of management and technology.

Examples here are organisational innovations such as lean production, lean management, and business process re–engineering which are based on research undertaken at the MIT and which have been taken up, developed further, and transferred

---

7 This is a main outcome of the research network (KISSIN) knowledge intensive services in Europe, financed under the TSER Framework Programme and co-ordinated by Prof. Peter Wood, University College London.

to other European countries by large, internationally oriented, knowledge intensive consulting firms.

The significance of small national or regional KIBS firms in innovation systems is mainly based on their *adaptation function*. Innovations are always associated with the communication and learning processes of firms. These are to a very large extent determined by social and cultural factors which differ not only nationally and locally but also at the level of the firm and which strongly influence the introduction of innovations. This culture and context linked knowledge, which is very difficult to codify, represents a major barrier to the internationalisation of consulting services. Small national or regional KIBS can adapt the knowledge of the innovative methods, that is to say, the specialist core of the consulting product that can be spread globally to the client specific situation by using their local knowledge and experience. It is they who, in a learning and interaction process, can use the knowledge about models and instruments as an element in the problem solving process.

4.6 Conclusions – Some Implications for Innovation Policy

In the process of globalisation of the economy, KIBS are developing increasingly into a non-institutional informal "knowledge transfer structure" and are thus an important element for systems of innovation. As has been shown, these services link the existing technological and management knowledge from separate disciplines, combine them in new ways, and adapt them to the appropriate business context. They contribute to change and innovation in a broad sense, not only through their innovative service products, but particularly through the indirect and feedback effects stemming from the introduction of these products. Because they are under continual pressure from intensive competition to build up new competences in their core businesses, they spread both collective know-how based on experience and best practices over different branches and by doing so, produce further new knowledge. They thus help with the transformation of knowledge into marketable products and contribute to the emergence, diffusion and adoption of technological, organisational, and social innovations. KIBS contribute to the acceleration of the innovation dynamic by means of these interdependent processes. The high degree of self organisation, which is one of their features, must therefore be seen as a strength with respect to innovative capacity. From this point of view, it would seem necessary to integrate this service segment into European, national, and regional innovation and development strategies.

Taking KIBS into consideration in innovation strategies, however, does not imply that there is an obligation to develop new branch strategies with reference to KIBS. Concepts for action whose goal is to support the KIBS function in the innovation system must be directed towards promoting the interaction and learning processes between the demand and supply sides. In addition, measures can only be effective if they take into account the specific national and regional socio–institutional context of the interaction. Political strategies for action thus require a decentralised, process oriented perspective which takes account of the high degree of self organisation in broad sections of the KIBS segment.

4.7   References

BADE, F.–J. (1990): Expansion und regionale Ausbreitung der Dienstleistungen. Eine empirische Analyse des Tertiärisierungsprozesses mit besonderer Berücksichtigung der Städte in Nordrhein–Westfalen, *ILS–Schriften*, 42, Dortmund.

BESSANT, J./RUSH, H. (1995): Building bridges for innovation: the role of consultants in technology transfer, *Research Policy*, 24, pp. 97-114.

BRACZYK, H.–J./COOKE, P./HEIDENREICH, M. (Eds.) (1998): *Regional Innovation Systems. The role of governance in a globalized world.* London: UCL Press.

BROUWER, R.E./KLEINKNECHT, T.A. (1997): Measuring the unmeasurable: a country's non-R&D expenditure on product and service innovation, *Research Policy*, 25, pp. 1235-1242.

COHEN, W.M./LEVINTHAL, D.A. (1990): Absorptive Capacity: a new perspective on learning and innovation, *Adminisrative Science Quarterly*, 35, pp. 128-152.

DEUTSCHES INSTITUT FÜR WIRTSCHAFTSFORSCHUNG (DIW) (1998): Innovationen im Dienstleistungssektor, *DIW-Wochenbericht*, 29/1998, pp. 519-533.

DEUTSCHES INSTITUT FÜR WIRTSCHAFTSFORSCHUNG (DIW) (1997): Dienstleistungsdynamik in der Europäischen Union uneinheitlich, *DIW-Wochenbericht*, 29/1998, pp. 519-533.

DOSI, G./FREEMANN, C./NELSON, R./SILVERBERG, G./SOETE, L. (Eds.) (1988): *Technical change and economic theory.* London: Pinter Publishers.

EDQUIST, C. (Ed.) (1997): *Systems of innovation: Technologies, institutions and organizations.* London: Pinter Publishers.

GAEBE, W./STRAMBACH, S. (Eds.) (1993): *Employment in business Related Services – An intercountry comparison of Germany, the United Kingdom, and France.* Report for the European Commission – DG V, Bruxelles.

HAUKNES, J. (1998): *Services in innovation – innovation in services*, SI4S final report, STEP Group, Oslo.

LICHT, G./HIPP, Ch./KUKUK, M./MÜNT, G. (1997): *Innovationen im Dienstleistungssektor: Empirische Befunde und wirtschaftspolitische Konsequenzen.* Schriftenreihe des ZEW, Band 24. Mannheim: Zentrum für Europäische Wirtschaftsforschung.

LUNDVALL, B.-Å./BORRAS, S. (1998): *The globalising learning economy: Implications for innovation policy.* European Commission, Science, Research and Development, TSER report, EUR 18307, Luxembourg.

LUNDVALL, B.-Å. (Ed.) (1992): *National systems of innovation. Towards a theory of innovation and interactive learning.* London: Pinter Publishers.

MOULEART, F./TÖDTLING, F. (Eds.) (1995): Geography of advanced producer services, *Progress in Planning*, 43, pp. 89-274.

NELSON, R. (Ed.) (1993): *National Innovation Systems.* New York: Oxford. University Press.

NOYELLE, T. (1996): The economic importance of professional services. In: OECD (Ed.): International trade in professional services. Assessing barriers and encouraging reforms, pp. 19-28, Paris.

O'FARRELL, P.N./WOOD, P.A./ZHENG, J. (1998): Regional influences on foreign market development by business service companies: elements of a strategic context explanation, *Regional Studies*, 32, pp. 31-48.

STARBUCK, W.H. (1992): Learning by knowledge-intensive firms, *Journal of Management Studies*, 29, p. 713-740.

STRAMBACH, S. (1999): Wissensintensive unternehmensorientierte Dienstleistungen im Innovationssystem von Baden-Württemberg – am Beispiel der Technischen Dienste, AKADEMIE FÜR TECHNIKFOLGENABSCHÄTZUNG IN BADEN-WÜRTTEMBERG (Ed.). *Arbeitsbericht 133*, Stuttgart.

STRAMBACH, S. (1997a): Wissensintensive unternehmensorientierte Dienstleistungen - ihre Bedeutung für die Innovations- und Wettbewerbsfähigkeit Deutschlands, DEUTSCHES INSTITUT FÜR WIRTSCHAFTSFORSCHUNG (DIW) (Ed.): *Vierteljahreshefte zur Wirtschaftforschung*, 66, pp. 230-242.

STRAMBACH, S. (1997b): *Knowledge-intensive services and innovation in Germany.* Final Report for the Commission of the EU - TSER Project No. SOE1-CT-95-1017 (unpublished).

STRAMBACH, S. (1995): *Wissensintensive unternehmensorientierte Dienstleistungen: Netzwerke und Interaktion.* Am Beispiel des Rhein-Neckar Raumes. Münster: Lit-Verlag (=Wirtschaftsgeographie Bd. 6).

TIDD, J./BESSANT, J./PAVITT, K. (1998): *Managing innovation: Integrating technological, market and organizational change.* Chichester, New York: Whiley & Sons

WOOD, P. (1997): *Knowledge-intensive services and innovation.* Final Report for the Commission of the EU - TSER Project No. SOE1-CT-95-1017 (unpublished).

WOOD, P. (1996a): Business services the management of change and regional development in the UK: a corporate client perspective, *Transactions of the Institute of British Geographers*, 21, pp. 644–655.

WOOD, P. (1996b): An 'Expert Labor' approach to business services change. In: *Papers in Regional Science, The Journal of the Regional Science Association International*, 75, pp. 325-349.

(Eurôu)

5. Institutions of Technological Infrastructure (ITI) and the Generation and Diffusion of Knowledge

D83 032

Antoine Bureth, Jean-Alain Héraud   038

5.1 Introduction

Institutions of Technological Infrastructure (ITI) are various forms of organisations that contribute to the actual or potential techno-economic development of the regions where they are located. Their types actually range from specific institutions created in order to fulfil the functions of technology transfer, R&D funding, consulting in innovative activities, etc., to various kinds of organisations like public research institutes and private firms which, while having other primary goals and rationale, play a role of ITI in their region to a certain extent and for a particular aspect of their activity. Starting from the now classical assumption that innovative development mainly results from interaction between several agents, and that proximity matters (more or less) in the building of such creative networks, ITIs as central actors of the "learning regions" are considered here.

Our aim is to exhibit a relevant representation of the cognitive interactions leading to collective learning, using the recent theoretical developments of the economics of knowledge, before presenting a typology of functions fulfilled by ITIs and studying the cognitive interactions at work in a concrete example of such an institution. Another question will be addressed through the analysis of the role of ITIs in a region: to what extent can we consider the region as a relevant level to describe the processes of technological development and innovation? ITIs can be viewed as central elements of the "Regional Systems of Innovation". However, all territories do not necessarily exhibit strong *systemic* characteristics, whatever their performances and originalities. Every actor of a territory actually plays roles in a whole set of systems on different geographical levels. Only "regional institutions", strictly speaking, are supposed to restrict their action to a given territory. Then the nature of creative interaction and collective learning, which is at the core of innovation and economic development, is a complex process where actors select their partners within the territory and/or elsewhere (with some specific biases towards the territory, that we still have to analyse) and follow a communication scheme around various forms of knowledge items (pure codified information, more or less tacit knowledge, individual and collective competencies, etc.).

For a better understanding of the role of the ITIs in such techno-economic developments, and for assessing the importance of the proximity or the regional "milieu", we first need to explain the various modes of generation of knowledge: the first section thus focuses on the learning processes. It describes how the properties of the learning processes (specialisation, uncertainty, path dependency,...) impose requirements in terms of proximity, not only from a geographical point of view, but also from an organisational point of view. Regarding proximity as an endogenous parameter and assuming that the interactions within a region are mainly concerned with knowledge, the second section proposes a structural representation of knowledge including the concept of competencies. Having described the concepts, in the third section the role and the functions of the ITIs are stressed. As an illustration, emphasis is put on the case of the CORTECHS procedure. The fourth section deals with the relevance of proximity and the possible meaning of the concept of a regional system of innovation.

5.2 An Endogenous Approach to the Local Dimension, Based on Learning Processes

A clear trend in the works dealing with the geography of innovation is to reverse the causality link between the economic organisation of an interacting group of agents and the territory in which they are installed. In the Weberian tradition, localisation is justified by the endowment of resources in a given physical area. Thus, the organisation of economic relations is strongly dependent on the territory, the latter being characterised by criteria such as transportation costs or the localisation of competitors.

When analysing the innovation processes, this relation can be turned around. Indeed, what matters is not only to structure economic relations around pre-existing resources, but much more to create resources within a pre-defined organisational frame. Those resources, mainly immaterial and informational, are generated through interactions and learning processes, and are the keystone of the innovative capabilities of the system under consideration. To this extent, the crucial aspect is how the agents achieve the exchange and benefit from the cross-fertilisation of their differentiated knowledge and competencies.

5.2.1 Knowledge as an Input/Output of the Learning Processes

A first step is to specify the representation of the learning processes which generate knowledge. Broadly speaking, learning can be defined as the discovery and the appropriation of a new piece of knowledge shaped by the learner in a suitable form. Following this definition, learning can be understood as the conjunction of four cognitive procedures (four processes of knowledge processing), i.e. creation, detection, memorisation and evaluation (Bureth 1994):

- creation is a goal oriented process of recombining existing pieces of knowledge, in order to generate novelty;
- detection is the capability to perceive and decode stimuli and messages existing in the environment of the learner;
- memorisation is the process through which knowledge is stored and retrieved;
- evaluation concerns the ability to select knowledge efficiently (knowledge has to be useful and usable).

A learning process relies on the combination of the four procedures, which are closely interwoven. Detection capabilities for instance depend on the memorised rules and knowledge, imply evaluation abilities in order to select the different pieces of knowledge, and are influenced by knowledge generated through the process of creation. Conversely, knowledge obtained by detection impacts on the way the three other procedures operate. Learning processes vary according to the relative importance of some procedures towards the others, and in the way they are articulated together: a student emphasises memorisation, a production operator favours the detection procedure in his learning process, a manager mainly uses the evaluation procedure, and a researcher should concentrate on his creation abilities.

Starting from this representation, the competencies of an economic agent can be defined as his/her capacity to co-ordinate the cognitive procedure with respect to the constraints and objectives imposed by his domain of activity. We distinguish competencies from knowledge, the former being the knowledge needed to process knowledge.[1] For instance, the doctor is not the one who knows the symptom, but the one who is able to recognise it in different contexts (which implies a capability to select, to interpret and to enrich the available information).

The distinction is important to grasp the phenomena of agglomeration and of valorisation of the proximity. Indeed, agents interact either to exploit complementarities in knowledge, or complementarities in competencies. Depending on the case, the requirements in terms of co-ordination differ. Exchanging elementary pieces of knowledge (information or messages) can be achieved through spot transactions,

---

[1] This argument is developed extensively in section 5.3.

sometimes under market rules. This does not mean that knowledge is a standard marketable good, but at least "commodification" of knowledge (Ancori *et al.* 2000) is possible. Now, depending on the context of the exchange, the degree of intimacy of the partners, the strategic content of the exchanged data, the frequency and the intensity of the interactions vary. Nevertheless, interacting on competencies implies a higher organisational cost: to exchange competencies, the protagonists have to integrate their cognitive procedures. It is time consuming, and it is realised within a specific organisational device, such as an agreement or an institutional procedure for instance. From this perspective, ITIs can play two different roles in the innovation process. On the on hand, they appear to be potential partners of the firms, providing knowledge. On the other hand, they can be involved in the design of the organisational frames within which partners will exchange competencies.

The proposed representation of the learning process is a way to separate *knowledge* as an output (obtained by the interplay of the cognitive procedures) from knowledge used as an input (*competencies*, i.e. the knowledge mobilised to co-ordinate the cognitive procedures in order to produce knowledge). The distinction is rather artificial - a single piece of knowledge can be successively an output and an input - but sheds light on different intervention possibilities in terms of public policy. Knowledge as an output raises the question of accessibility: if knowledge cannot be produced directly by an actor, how will he/she obtain it? In that case, the means used for storing and supporting knowledge are mainly under discussion. Knowledge as an input induces another type of questioning: if an actor does not hold the right kind of competencies, how to assimilate it? In the latter case, the emphasis is put on the processes which make it possible to generate and to transfer knowledge. It raises the questions of the irreversibility and the uncertainty attached to the learning processes, which are developed in the following lines.

5.2.2   Proximity and the Management of Irreversibility and Uncertainty

Basically, agents reinforce and enrich their knowledge along two lines of learning.

First, new knowledge is acquired *as a joint-product of the production activities*. It refers to "learning by doing" described by Arrow, or Rosenberg's concept of "learning by using". Such learning processes are dependent on the competencies of the agent or the firm: the volume, the value, the specificity, the degree of codification, in other words, the properties of the acquired knowledge rely on the cognitive abilities of the learner. Furthermore, those individual learning processes (expressing experience effects) lead to specialisation. Learning is localised (Stiglitz 1987) and is driven by procedures and routines self-reinforcing over time. As far as learning includes a synthesis and a selection of knowledge, it generates irreversibility by framing possible future options. Nevertheless, beside the property of path dependency, the cumulativeness of learning allows increasing returns and is a necessary condition for the development of novelty and innovation.

This contrasted picture illustrates the main issue of individual learning processes: there is at the same time a need to stabilise the practices and knowledge and a need to preserve the capability to explore new domains. Competencies, seen as the instruments of the management of the learning processes, aim at achieving this trade-off between irreversibility and flexibility of the learning trajectories (Bureth 1994). The assessment of public intervention in terms of innovative capabilities should take into account this paradoxical requirement. Especially ITIs have to sustain what James March called *exploitation trajectories*, but have also to keep the potentialities of exploration in new fields of activities.

Secondly, knowledge can be acquired *through interactions* (co-operation, partnerships, etc.). Such an operation can be compared to a risky investment, and requires resources which will be lost if the co-operation aborts. Furthermore, learning by interacting also generates instability: co-operation raises knowledge about the partner, about the domain of co-operation, about the alternative opportunities of development, and a part of this knowledge can be applied outside of the relationship. Hence there is an incentive for one of the partners to break the relation in order to exploit the advantages obtained through the co-operation alone (Bureth *et al.* 1997). Learning by interacting must obviously respect the balance between irreversibility and flexibility described previously. But in order to develop the right competencies, there is an additional constraint: interaction has to be maintained over time. The determinant factor - on which a public co-ordinator can play a role - is the relational stability.

Our aim in this contribution is precisely to study to what extent being part of a localised system can improve the learning facilities and/or potentialities. In our view, proximity is a necessary condition to implement efficient learning processes. However, the concept does not only refer to a physical localisation: it includes an institutional dimension and evolves over time. As a variable, proximity can be modified by the action of a public co-ordinator and constitutes a central objective for the scope of intervention of the ITIs.

Geographical proximity appears to be the most "natural" requirement in the existence of an organised group of firms. It favours mutual knowledge and direct adjustment between the agents, creates the conditions for a common history. However, those advantages do not systematically grant the creation and the valorisation of knowledge within the geographical space: in other words, the innovative capabilities of a local system do not rest in the density of the agglomeration. And the notion of geographical territory imposes *ex ante* an analytical frame which does not necessarily fit the organisational borders drawn by the collective learning processes. In our understanding of the innovative processes, space is structured by the mechanisms driving the evolution and the exchanges of knowledge and competencies. The organisation of a set of firms will be delimited and localised following the sharing of technological and organisational knowledge. The "local" dimension thus becomes an endogenous variable and the territory is assessed *ex post*, looking at the way knowledge is produced and diffused.

On the one hand, the territory is an economic space within which the knowledge produced through learning processes can be valorised: explicit and/or implicit rules define the property rights and the conditions of accessibility, use and diffusion of knowledge. On the other hand, the territory provides facilities and gathers the resources needed to produce valuable knowledge: the existence of organisational and institutional devices makes the sharing of competencies easier.

Following this perspective, two main features can be attached to the concept of proximity. First, proximity is reinforced by the existence of a common knowledge base, to which agents participate and refer. Beyond facilitating the acquisition of knowledge, the common knowledge base is a crucial element of a local system of innovation insofar as it homogenises the individual learning processes and it reduces the diversity among the competencies of the agents. It reduces the cultural distance (Lundvall 1992), smoothes the differences between the bases of experience (Guilhon/Gianfaldoni 1990). Consequently, communication costs due to the differentiation of knowledge and competencies decrease and interactive learning processes are easier to implement.

Secondly, proximity is dependent on the degree of uncertainty related to collective learning processes. Indeed, interaction processes include a strong uncertainty: the result of the co-operation is not predictable, there is a lack of knowledge about the behaviour and the potentialities of the partner, the sharing of the rent created by the co-operation cannot be defined ex ante, etc. Here geographical proximity helps once again to limit uncertainty. Personal and informal relations, or reputation effects will impact on the co-ordination mechanisms and will stabilise behaviours. But the reduction of uncertainty also relies on the implementation of organisational devices. Fairs, meetings, inter-firms training are occasions to multiply inter-personal contacts. In a more goal-oriented way, the intervention of an institutional actor also helps to manage uncertainty. Norms and certification provide an illustration of it: the ISO 9000 expresses basically a competitive advantage on the market. But the validation of the productive procedures also signals that the firm is a reliable partner. To the same extent, public agencies assume an intermediation role in co-operation between research and industry by managing the legal and financial aspects of the contracts. As expressed by Kofman/Senge (1993), "learning occurs between a fear and a need". Rules and procedures reinforcing the institutional and organisational proximity make it possible to reduce the fear, and may help to define the needs better.

Looking to the learning processes raises a certain number of questions concerning the role of the ITIs. First, they can impact on the individual learning capabilities of the actors, by providing knowledge or expertise. Secondly, they can improve the collective learning processes, by managing the innovative network or diffusing common languages or codes. Thirdly, they can "reveal" some needs incompletely expressed both on the demand or on the supply side. But before considering those different functions in detail, it is necessary to describe the different forms of knowledge mobilised in innovative activities.

5.3   The Structure of Knowledge

What comes out of the previous section is that proximity is an endogenous variable, depending on the conditions of learning. Different forms of proximity (geographical, but also organisational or institutional) have an impact on the management of the diversity of knowledge and on the management of uncertainty. In discussing the learning processes, we have introduced the distinction between knowledge and competence, the latter being the knowledge required in order to produce, acquire and exchange knowledge. This section will be devoted to the exploration of this point in depth and will provide a representation of knowledge as a structure, which can be broken up into complementary layers.[2] Categorisations such as the one proposed by Lundvall (1998) is not sufficient for understanding what the innovative capabilities rely on: in a world in which the tremendous development of the new technologies of information and communication increasingly facilitates the access to knowledge, what becomes crucial is the cognitive resources to deal with knowledge. That is why we prefer a functional decomposition of knowledge[3] rather than an identification of the "products space" of knowledge. We distinguish the "crude knowledge" from the knowledge about how to use, to transmit and to manage

---

2 At this point, in line with what was expressed by Machlup/Saviotti in a recent article (1998) summarised some of the main features of knowledge: *"Knowledge establishes generalizations and correlations between variables. Knowledge is therefore a correlational structure. The extent to such correlation is not infinite. Each piece of knowledge (e.g. a theory) establishes correlations over some variables and over particular ranges of their values. As a consequence, knowledge has a local character ... The degree of such local character can be measured by the span of a given piece of knowledge, that is, by the number of variables and by the amplitude of the of the range over which correlation is provided. General theories will have a greater span than a very specific piece of applied knowledge ... Particular pieces of information can be understood only in the context of a given type of knowledge. New knowledge, for example relative to radical innovation, creates new information. However, this information can only be understood and used by those who possess the new knowledge. In other words, knowledge has also a retrieval/interpretative function. In summary, knowledge is a correlational and a retrieval/interpretative structure, and it has a local character".*

3 For more details see Ancori *et al.* (2000)

knowledge, the last three categories (ranked in order of increasing complexity) constituting the competencies are described in section 5.2.

a) As soon as stimuli (emitted by nature) and messages (emitted by a human being) are perceived by an agent, both constitute the first layer of the structure of her/his knowledge we call *"crude knowledge"* - i. e. information before the receiver builds a meaning, as seen by Shannon/Weaver (1949). This layer of knowledge gathers elements such as plain facts, messages and intuitions (which are kinds of "reflexive" messages). At this elementary step we consider these elements as autonomous pieces of knowledge. The point is not to see if they have been, or how they will be processed, interpreted or attached to a specific context: if we think of the learning of a foreign language, the emphasis is put on "grasping the word" before "getting the meaning". This is obviously a rather artificial perspective, but useful to underline a first issue: when considering crude knowledge, what matters is the volume of knowledge to be perceived compared to the (scarce) cognitive resources of the agents. For instance, the learning processes in collecting crude knowledge elements will be different if they are highly codified or if their sources are localised in time and/or space because of organisational boundaries or rules.

b) The second layer of the structure of knowledge is the knowledge about *how to use (crude) knowledge*. This first type of competency makes it possible to articulate a new piece of stimulus/message with the previous stock of accumulated knowledge, and to relate action and decision making to crude knowledge. Once a stimulus or a message has been perceived, the agent has to resolve the equivocability of this new piece of knowledge. This is done by a backward process through which the new knowledge is confronted and articulated with previous experiences.[4] Weick (1979) distinguishes three processes: the enactment process, in which the "knower" *"constructs, rearranges, singles out and demolishes many objective features of his/her surroundings"*; the selection process, which leads to identifying, rightly or wrongly, the set of causal relations; and finally the process of retention, which is the memorisation of the successful sense-making.

Knowledge thus appears *"... as a map of relationships between events and actions, that can be retrieved and superimposed on subsequent activities"* (Choo 1996). However, those links can be more or less loose. In some cases, the perceived stimulus automatically releases a well defined action, as for the Pavlovian reflex. It is the case when previous learning processes have produced routines ; and the more the routines are achieved and adapted, the less understanding is required to justify

---

[4] We refer to Weick's representation (1979) of the organisational sense-making process albeit a cognitivist would certainly fully disagree. But for economic purposes it seems also to be applicable to individuals.

the undertaken action.[5] Conversely, in some situations, the crude knowledge has first to be articulated with previous accumulated knowledge which must be retrieved and eventually re-organised. Then, if the consequences are not clearly identified, action will be undertaken step by step, through trial and error processes, in order to collect more knowledge ("crude knowledge") about its reliability.

c) The third layer (*competencies in the transmission of knowledge*) is related to the exchanges and interactions an agent has to conduct in terms of knowledge. This category mainly covers codes, languages and models, which are basically tools used to exchange knowledge. Thus, knowing how to transmit knowledge obviously implies the mastery of codes and/or languages. Furthermore, it includes knowledge about the modes of conversion of knowledge (Nonaka 1994), which are the ways through which individual knowledge becomes collective (and reciprocally), tacit knowledge becomes explicit (and reciprocally), two pieces of knowledge merge, one piece of knowledge splits into two, etc. There is at this stage an implicit strategic dimension insofar as the adopted codes, languages and modes of conversion adopted to diffuse knowledge are not neutral in the transmission. They intrinsically include a representation of the world and mobilise different amounts of cognitive resources, both for the sender and for the receiver. Furthermore, the strategic dimension also includes the question of the incentives to reveal or to keep secret the code used to process knowledge. The decisions which take those aspects into account rely on competencies described in the next category.

d) The fourth and final layer of knowledge is the *knowledge about how to manage knowledge*. This category corresponds to higher cognitive functions, which shape the three previously described categories.[6] The management of knowledge supposes the ability to modify even drastically the context of production and exchange of knowledge. It encompasses thus knowledge about when, where and how to "find" relevant crude knowledge. We mean here knowledge about the localisation of knowledge as well as knowledge about the incentive mechanisms which will bring another agent to deliver messages about what he knows. It also includes knowledge related to the enactment, the selection and the retention modes mentioned in the second layer. Are those processes efficient? Should they be improved (and how)? Are they too costly, etc.? Knowledge is also needed for the management of communication. The choice of the conversion modes or of the used codes will occur according to the use of the transmitted knowledge the other agents could make. The same type of choices can insure a relative control of the appropriability of codified knowledge. Finally, the management of knowledge also depends on the degree of interaction between the different sub-categories of knowledge. Under such assump-

---

5   March/Simon (1993) defined the concept of performance programmes, in which a pre-defined action programme allows the reduction of the requirement for cognitive resources.

6   The learning processes at stake come close to the Bateson (1972) third type learning or the deutero-learning characterised by Argyris/Schön (1978).

tions, the *knowledge to manage knowledge* becomes a crucial and strategic resource, able to create decisive advantages in favour of its holder(s).

One of the main motivations of this representation is to avoid associating strictly one type of knowledge (a product of knowledge) with one knowledge based activity - science deals with *know why*, industry focuses on *know how*, etc. The questions that matter here are the following: how is knowledge produced? Is it incorporated in human capital or stored in physical artefacts? Is the use of knowledge submitted to strict rules? Is there room for ambiguity and superstitious learning? What is the degree of control exerted on the mechanisms of diffusion of knowledge? All of these questions can be addressed to different activities using common criteria in terms of learning processes.

Looking at the production of knowledge, processes of *learning by doing* and *learning by using* are mainly at stake. A first aspect is the relevance of the crude knowledge hold by the agents and their competencies in using this knowledge. The more the link between actions and stimuli/messages is clearly established, the more the activities (scientific research, industrial production, service activities) will be routinised and stable. The irreversibility thus generated should be counterbalanced by the knowledge about how to manage knowledge. This upper level should include expertise capabilities mobilised to avoid an inefficient lock-in in the production of knowledge. In other words, the knowledge about how to manage knowledge should, if needed, interact with the knowledge about how to use knowledge, and subsequently modify the accumulation of crude knowledge.

When learning is done collectively, codes and languages are obviously required to communicate. And once again, there are interactions between the different categories of knowledge. Exchanging crude knowledge is not sufficient. The interpretation capabilities - i.e. the knowledge about how to use knowledge - also have to be shared, through models and representations. Furthermore, the knowledge about how to manage knowledge is needed to choose appropriate codes enabling control of the degree of diffusion of knowledge.

Another point can be raised concerning the exchanges of knowledge. Knowledge does not necessarily have to be transmitted in order to be processed collectively. Part of the crude knowledge is often specific, tacit, or held by individuals. The same properties can be attributed to contextual know how, belonging to the second category. In this case, collective learning processes are still possible, even if the knowledge is personal, tacit and "sticky" (von Hippel 1993). Nevertheless, co-ordination processes are required: like in the functioning of a team, the collective learning process is dedicated to a common objective, but relies on decentralised stocks of individual knowledge. Co-ordination of the learning processes is thus a substitute for the codification of the knowledge itself. A symmetric situation can be about the knowledge about how to manage knowledge. For such a complex form of knowledge, transmission can be impossible even if the codes to communicate it exist). It is for instance the case if we look at the knowledge held by an expert. Potentially, part of his/her knowledge can be formalised and transmitted, but in practice it is almost impossible. The explanation we propose is that the knowledge about how to manage knowledge remains tacit. Like the carpenter of Polanyi, the expert knows, thanks to a long practice, which knowledge has to be mobilised and which knowledge has to be left in the background in order to act or to learn properly. The difference with the previous situation is that under certain conditions (common history, shared experiences, social or organisational frame), such a form of tacit knowledge can be communicated and shared among the members of a community. And its possession determines the insiders and the outsiders of the community. In other words, what matters in this case is the history of the construction of the tacit knowledge. To manage tacit knowledge implies modifying the time dimension (by intensifying the interactions for instance) more than transforming the knowledge itself.

5.4   Managing the Institutional Landscape of the Regions: the Functional Approach of the ITIs

The functional approach addresses the ITI not as a fixed *ad hoc* institution, but in accordance with the functions fulfilled. In some cases, it is an organisation whose main function is officially to organise the "system of innovation" (within a state, a region or any sort of territory) which generally means linking actors together or connecting them to any global source of information. In other cases, the meaning of ITI is not that of an "institution" in the ordinary sense: a large firm can play the role of provider of information and technical norms, be a show-room for specific technologies, introduce new vocational competencies in the region, etc. In other words, the important aspect is that an ITI is involved in the organisation of the collective structure of knowledge (cf. Figure 5-1).

Considering knowledge as the keystone of the activities of the ITIs has an important consequence. As already mentioned in section 5.2 (proximity is endogenous), the relation to the territory varies. The ITI is a component of the regional landscape, but it is not necessarily devoted to the territory like an organ would be in a biological organism. We should not speak of "regional system" in the sense of an organism, but rather as an ecological system, which means in fact a set of overlapping and hierarchical systems.

Figure 5-1:   **Possible actions of ITI**

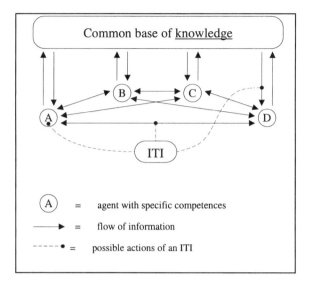

Source: Koschatzky/Héraud 1996: 3-6

Institutions like CRITT in France (regional centres for innovation and technology transfer) are mainly territory-oriented since they are funded by the region and are supposed to respond to the regional firms' technological demand , or to "reveal" such a demand, and/or to valorise the competencies of scientific/technological institutions located in the region. Even in such cases sometimes the reality lies far from theory: in Alsace for instance, the regional administration (Conseil Régional) has long complained that the specialised CRITTs (which are co-financed by the Conseil Régional and the central State) deal more often with firms outside the region due to the fact that there is more technology demand outside of Alsace.

Furthermore, public research labs (university institutes or regional establishments of national research organisms) also play an important role in the regional landscape, without being devoted to the region. The latter are main components of the national system of innovation and are linked to the community of science world-wide; if they efficiently work in such global systems they can be all the more useful in the regional landscape (although not necessarily), but regional diffusion of S&T information is not their first mission.

These few examples show the usefulness of approaching ITIs in terms of the cognitive functions fulfilled within the territory, instead of trying to build typologies starting from their legal definition and organisational structure. The same empirical and analytical approach is to be adopted for the spatial dimension: the territory should not be viewed as a concept *per se,* an exogenous variable, but in a more constructivist way, in the sense that every actor actually designs its own relevant territory (and even several territories - one for each main function).

We will start here with a presentation of the main categories of ITIs and cognitive functions ensured. Since the institutional landscape is not composed only of ITIs but also of procedures and because varied aspects of regional settings must be taken into account, we will then present a case study in the French Region Alsace: the CORTECHS procedure involving several actors and fulfilling several functions. We will conclude by addressing the question of the relevance of geographical delimitations, i.e. asking "if there is something like a regional system of innovation", and assessing the concept of proximity for innovative networks.

5.4.1   Defining and Characterising ITIs

In previous works (cf. Koschatzky/Héraud 1996)[7] we considered that Institutions of Technological Infrastructure could be defined as entities:

- which have a legal identity (in public or private law);

- which are located in a specified region and have potential technological impact within the territory - e.g. influencing the innovation capacities of firms located here;

- the activities of which provide the input for research and innovation of enterprises by fulfilling the following functions: managing the knowledge base, improving interactions between enterprises or providing expertise knowledge.

As for the preliminary definition in terms of legal status, we must first consider that some of the ITIs are institutions *stricto sensu*. Among them, the most important are:

- Departments, institutes, or research teams of public organisms (university, engineering schools, public research centres) or semi-public entities linked to industrial branches, chambers of commerce, etc.

- Education and training institutions acting as diffusion centres or supporting partners in certain procedures (like CORTECHS, as we will see in 5.4). For this function they operate more with *knowledge about how to manage knowledge*.

- Organisms officially in charge of the function of technology transfer (like CRITT). They must have strong competencies in the transmission of knowledge.

---

7   On behalf of Eurostat, Fraunhofer Institute for Systems and Innovation Research (FhG-ISI, Karlsruhe) and Bureau d'Economie Théorique et Appliquée (BETA, University of Strasbourg) carried out a feasibility study on the definition and statistical measurement of the Institutions of Technological Infrastructure (ITI). The objective of the study was to derive methodological recommendations for the statistical measurement of ITI at national and a regional level and to prove their practical feasibility.

Other actors fulfil a role of ITI. Private firms have already been cited. One can first consider any manufacturing corporation that plays a leading role in the local industrial fabric. A typical example would be a large "prime contractor" dealing with a whole set of sub-contractors and suppliers. In a way such a firm functions as one entry point of the regional landscape, opening up the local sub-system to upper-level systems. Conversely, smaller firms provide specific innovative factors within the regional environment, playing a crucial role for its global attractivity. Some of them are manufacturing SMEs, specialised sub-contractors of larger firms, contributing notably to the global competitivity of the latter. It has been shown in previous studies in Alsace (Héraud/Nanopoulos 1994a) that such specialised suppliers interact to a large extent with their clients for improvement and innovation in the final product.

Another type of small firms that clearly play a role of ITI is the category "Knowledge Intensive Business Services" (KIBS) examined in other parts of this volume (see the contributions of Strambach and Muller). They can be described as firms performing, mainly for other firms, services encompassing a high intellectual added-value. Thus, KIBS correspond broadly to "consultancy services".[8] They are perhaps closer to the definition of institutions than the specialised manufacturing SMEs. They contribute to various cognitive functions in the regional institutional landscape, mostly enhancing the capabilities of firms through business consultancy and in all the specialised areas of management (marketing, accounting and legal consultancy, information and communication systems, intellectual property, etc.). The impact on firms' innovation is sometimes direct, as in the case of external R&D performance, but indirect support is probably the major contribution of KIBS to regional SMEs' innovation, since the main obstacle (at least in a region like Alsace, where a strong manufacturing tradition exists) seems to be the lack of general business competencies and of strategic information, more than the lack of technological capabilities. To sum up, KIBS can complement the SMEs for some vital internal functions, and their (inter-)active presence in the regional environment makes a considerable difference in terms of potential development and even survival of the smaller units in a world of increasing globalisation.[9]

These examples have shown the variety of the roles fulfilled by ITIs, following the functional definition we propose. One could restrict the concept of ITI to the sole category of institutions like centres of technology diffusion and public labs, but in that case, a large part of the actual processes leading to innovation would be implicitly ignored.

---

8 A detailed description of KIBS can be found in Miles *et al.* (1994).

9 Their specific contribution in terms of knowledge interactions make them unavoidable for regional development, especially for less advanced regions and for regions suffering from a weak innovation-supporting infrastructure (cf. Muller 1999).

5.4.2   The General Functions of ITIs

The varied functions fulfilled by ITIs can be characterised in terms of the general theory of information and knowledge which have been developed in the first sections of the text. For instance, the mode of collective learning occurring between KIBS and firms generally has a strong tacit knowledge content, whereas knowledge exchanged with more "institutional" ITIs is often codified (typically, the access to scientific knowledge by interacting with a university lab, or to the state of the art of a given technology through a specialised "broker" like a CRITT). We now need to proceed to a more detailed description of the informational nature of interaction with ITIs.

The third point of the general definition proposed concerns the three main types of cognitive functions of ITIs in a system of innovation: (i) managing the knowledge base; (ii) improving interactions between enterprises; (iii) providing expertise knowledge.

1. Managing the knowledge base:

Basically, the different functions mentioned here refer to the "commodification" of knowledge: ITIs have to ensure *accessibility to crude knowledge*, and they help to diffuse the *knowledge about how to use knowledge*.

1.1 "producing scientific and technological knowledge"	New scientific and technological knowledge is the output of basic and applied research. It results in an organised set of codified information. The typical institution here is the university or any public research organisation using as main diffusion medium scientific publications.
1.2 "educating"	Education as a form of knowledge-building is a critical input to the innovation process. Higher education and research organisations mainly perform this function.
1.3 "informing"	Information is made available to all innovative agents without discrimination. The diffusion of scientific and technological information uses tools such as data bases, synthesis publications, libraries. Mostly, added value is due to the concentration of information at a given point in the network.
1.4 "demonstrating"	Diffusion of knowledge and know-how can take place through demonstration activity, either by offering facilities to agents for testing products and processes or by acting as a lead user in order to explore the characteristics of technologies and their potential applications

2. Improving interactions between enterprises:

The second set of functions deals with the relational stability within an innovative system or region. The actions of the ITIs are mainly oriented in order to facilitate the *exchanges of knowledge between partners.*

2.1 "organising"	Promotion of networks of innovative agents: organisation of meetings, business fairs and exhibitions.
2.2 "financing"	Providing financial resources to innovative agents can be varied and very complex. In this case, financing interactions is used as an incentive mechanism to improve relations between agents. Related to this are services such as finding partners, helping to set up a business plan and assistance in innovation management.

3. Providing expertise knowledge:

ITIs intervene here in the domain of *the knowledge about how to manage knowledge.* The learning processes are complex, specific, and strategic and they concern the core competencies of the firms.

3.1 "training"	While the education system aims at satisfying the collective needs of the whole innovation system, training adds specific competencies - orientated towards action and defined goals - taking into account the specific needs of individual actors.
3.2 "consulting"	Supporting and assisting the innovation process means providing complementary services such as consultancy on general strategy or on marketing, value analysis, legal advice, etc.
3.3 "validating"	Validating comprises any action leading to general recognition of a given technology (product or process).
3.4 "appropriating"	Setting up procedures defining property rights necessary to protect any innovative action: patents, copyrights, trademarks, etc.
3.5 "financing"	Here, financing means providing additional external resources for material or immaterial investment to the individual agent.

It is possible to come back to a typology of ITIs after reviewing those generic functions. We can at least consider three main categories:

a) Institutions belonging mainly (but not necessarily) to the public sector, whose primary function is the production and diffusion of knowledge in the sense of academic activities. The "academic world" corresponds broadly to higher education and research organisations located in a region. Such actors of the regional scene fulfil specific functions in the environment of firms: they support research and innovation activities through their own interaction with the general scientific and technological knowledge base; they sometimes contribute to circulating information and developing interactions between firms; and they provide potential facilities to the firms for improving or developing internal knowledge (expertise). Such a networking activity is interpreted in terms of economics of knowledge (cf. Héraud/Muller 1998: 2-4), as the ability:

- to collect and understand *information* flows;

- to develop *knowledge* bases, from these external information flows as well as from internal learning processes;

- to apply knowledge to problem solving activities, *i.e.* to build specific *competencies*.

In other words, they are mainly involved in the production of crude knowledge and of the knowledge about how to use knowledge. Considering those forms of knowledge, the important requirements are the physical and the financial means to sustain learning processes. This first category of ITIs needs tools for improving the conditions for the production of knowledge.

b) Support organisations of public or semi-public nature, providing assistance to firms. Support organisations include entities linked to chambers of commerce and sectoral organisations, national or regional agencies and associations. They mainly provide assistance to firms. The function is to find a good partner to resolve the problems of firms. They also try to help firms to express their needs. Some of them propose financial support to firms for implementing their projects. These institutions often work in networks: several support organisations (sometimes on different levels: local, regional, national) are associated in support procedures concerning a given firm. With respect to our representation of the structure of knowledge, they focus on the intermediary categories: knowledge about how to use and how to transmit knowledge. In that perspective, they will impact on the "local" development insofar as they achieve a real co-ordination power. They sufficiently master the network and its interfaces in order to manage the behaviours of the actors. Therefore what matters is more the "credibility" of the ITI than its financial resources. ITIs of that type, more than the "academic" ITIs, seem to be very specific to both national and regional institutional settings.

c) Private services such as consultants (KIBS). They clearly perform several of the functions previously exposed for their clients. Moreover, the contribution of KIBS to the innovation capacities of national and regional systems takes direct and indirect forms, as emphasised by Strambach in this volume: on the one hand (supply side) they contribute to the competitiveness of national and regional economies through their own innovation process; on the other hand (demand side) their cognitive interaction with clients contributes to revealing the potential needs of firms. In terms of the theory of knowledge, we would interpret the indirect action as a learning process on a higher level, leading to the production of knowledge about how to manage knowledge. An important feature of this form of activities is the need for time. To impact efficiently, KIBS have to interact with firms in the long run, as far as their activities imply technological but also organisational and even social changes. To this extent, activities of consultancy should not only be ruled by market mechanisms, but should be included as a specific tool in the public policies.

The next section will show a similar cognitive scheme (implementing a "learning to learn" process) in the case of a policy setting including academic institutions.

5.4.3   ITIs in Context: the Case of the CORTECHS Procedure

As in an eco-system, ITIs and other actors do not work alone. Even networking facilities are sometimes linked together in more global networks (in fact, networks of higher level). The general context of working is strongly structured by the policy settings, notably procedures designed at the national level in the case of France. It happens that certain significant experiences are tested at a regional level, then proposed nationally as a new model of organisation. One example is the CORTECHS procedure (Héraud/Kern 1997) initially tested in Alsace (1988-89), that proved to be ideally adapted in a region characterised by few industrial giants of the high tech sectors, but many active SMEs in varied "medium technology" sectors. A large number of French regions have now adopted the model, but Alsace is still ahead for the number of contracts of that type and there is a regional consensus to consider the procedure as one of the most successful policy tools for supporting innovation.

CORTECHS is a contractual procedure ("Convention de Recherche") between an SME and an ITI, involving a younger technician ("TECHnicien Supérieur"), recently graduated, who will be responsible of an innovative project within the firm. The ITI is academic (technical school or college, possibly a university lab) or sometimes a centre of technology transfer (CRITT). It is called "the competence centre" and supports the project in different ways: complementary training for the technician, general information, possibly the supply of technical facilities. A team of specialists (*"l'équipe-projet"*) is also associated to the technician and the people from the firm for supervising the innovation (or quality management) project. The firm pays about half of the cost of hiring the younger technician, the state and the region fund the rest. State and Region share the public cost of the procedure, within the framework of the global negotiation called *"Contrat de Plan Etat-Région"*. Specialised services of the State located in the region help with the administrative and technical engineering of the procedure.

It can be seen that a lot of actors are involved in each of those contracts. There are also many individual and collective learning experiences around the project: firstly, it is a very stimulating training experience for the technician after his or her education period; the firm gets an opportunity to test a risky innovative strategy (it is generally their first experience) at lower cost and with the help of specialists; the other actors of the procedure also learn through the project and can possibly apply that knowledge in further circumstances. The statistics prove that a large proportion of firms having innovated thanks to a CORTECHS, try again in the following years (with or without a new CORTECHS).

In terms of knowledge flows, the following scheme can be observed:

- The common knowledge base is used by the ITI (competence centre) to train the technician and to complement the knowledge about the firm.

- The technician helps to "translate" a lot of codified information learned at school or provided by the ITI into practical (largely tacit) knowledge usable for the firm. He accumulates his/her own competencies (codified and tacit).

- The firm learns about codes: how to directly or indirectly use the common knowledge base, and what the value of codifying (at least partially) the routines is in order to improve or replace them (innovation); new collective tacit knowledge is created around the innovation; knowledge about external actors is developed; superior level learning is achieved because learning how to innovate means learning to learn.

- Part of the learning around the project will be transferred indirectly to the rest of the innovation system through the experiences of the technician (if he/she moves to another firm after the stay) and of the ITI and other experts.

It is interesting to stress a last point concerning the success of the CORTECHS procedure. Experts of the system explain that the qualification of the young technician is the ideal level for the job. Most of the SMEs who benefited from the procedure would never have accepted a contract of the same type with an engineer. It appears that cultural and sociological parameters put a strong constraint on the networking possibilities. This point applies to the cognitive scene too!

5.4.4   Proximity and Regional Systems

Now the central question of this paper will be addressed: the spatial characterisation of the interactive learning linked to innovation in a region. The point is complex since the geographical projection of the varied networks involved in the innovation processes rarely coincide with any unique and coherent space. We are already confronted with this problem when considering general economic variables. We know if a precise establishment of a firm is located within the administrative limits of a region; we can usually assign most of the labour force to some regional administrative unit or statistical area; but for the rest of the economic and managerial characteristics, the situation is rather equivocal. For instance, what is the "regionality" of the capital invested in this establishment? Space appears to be a strongly multidimensional variable if we list the different business functions of the firm: its specific strategic field is, possibly, local/regional for the human resource and the sub-contractors, national for a lot of institutional links, European for selling markets, and world-wide for equipment. We must consider territories in a "constructivist" way, every agent designing its own set of relevant spaces.

In terms of innovative activities and learning networks, we find at least the same level of complexity. The model case of "districts" seems to be the exception rather than the rule. In Alsace, we tested, some years ago, the geographical extension of the innovative networks (external sources of innovative ideas as well as real networks and partnerships) on a sample of 200 Alsatian firms (establishments) (Héraud/Nanopoulos 1994b), and compared these geographical structures with the corresponding situation in Baden-Württemberg (Lake Constance area) (Hahn *et al.* 1995).

The main result of our study on 1400 strategic partners or external sources of innovation in the Alsatian industry (observed during the early 1990s) is that only one quarter of the links remain within the limits of the region. When restricting the sample to the SMEs, that proportion gets higher, but remains weak: 26 % of the formal strategic agreements leading to innovation concern Alsatian partners; the figure is 27 % for the other external sources of innovative ideas. By types of partners or sources, the results confirm the regular theoretical assumptions, showing a rather strong influence of the clients and the suppliers in the innovative learning processes, and a significant role of hierarchical transfers (particularly reflecting the very high rate of direct foreign investment of international groups in Alsace). The dominant geographical patterns of these innovation links is, in relative terms: France for the clients, Germany and the rest of the world for the equipment suppliers and the cross-border "Upper Rhine" region (including parts of Baden-Württemberg and North-Western Switzerland) for transfers from the parent-company. There is only one type of actor to be found systematically in the region: the ITIs.

Empirical findings exhibit a paradoxical situation: the innovative networking with institutional partners (ITIs such as chambers of commerce, regional administrations, sectoral research centres, universities, etc.) has a minor impact in relative terms; but when a firm needs such a contact, it looks *a priori* for a partner within the region. For example, among the small set of 35 innovations resulting from contacts with academic institutes in our sample, 19 concerned the Alsatian universities and public research labs, 12 were found in the rest of France and not one in the German and Swiss "Upper-Rhine" areas. To sum up, regional scientific institutions do not play a leading role for regional innovation at industry's level (although they are rather well developed and constitute an important piece of the French national system of science), but when industrial demand exists for such competencies, regional firms do not look beyond the national or regional S&T systems.

Private consultancy firms are the only non-institutional partners that play a specific role within the regional space: again, their weight is modest in the total of the innovation network, but when a link exists, it is generally with a regional unit of that type. The fact confirms our analytical choice to include KIBS in the category of ITI.

A last empirical finding must be stressed, concerning the role of proximity. For a whole set of socio-economic characteristics, it is possible to show the relatively strong convergence and coherence of cross-border regions like the Upper-Rhine area. A lot of goods, services, workers etc. flow through the state borders, and the industrial specialisation is relatively similar in all parts of the area. We expected then also to find a significant proportion of innovative collaborations, extending across the Rhine (and across Lake Constance); but the results proved the contrary. French, German and Swiss firms have not developed a significant innovative *milieu* (coherent space of collective learning) for R&D and innovation. The national systems of innovation are powerful realities, that can find regional translations, but very few international crossings. Proximity is not mainly geographical; it is linked to cultural and institutional settings (see also Koschatzky 2000).

5.5   Conclusion

ITIs are essential actors of *learning regions*, characterising the potential of the involved *territories*. The "territory", progressively designed as an interdisciplinary concept among economists, geographers, sociologists and regional planners, must be understood as a spatialised human organisation of medium size (local/regional), largely self-organised, mainly structured through the networking activity of actors localised within the area. Furthermore, it is a multi-dimensional and dynamic concept. In the cognitive approach we have tried here to develop around the definition and functional description of ITI, one important aspect has been underlined: the openness of the systems at hand.

Innovation processes through interactive learning are simultaneously localised (to the extent that they are at the basis of the concept of territory) and strongly interlinked at various geographical levels. Proximity matters, but it does not imply the existence of exclusive system links. Almost by definition, knowledge creation is a product of open systems. In this framework, ITIs are core actors whose efficiency in a given territory is certainly not positively correlated with territorial exclusivity. We must add that some of the ITIs are not institutions in the narrow sense, but actors whose primary activity is different. To summarise, we have shown that proximity as a necessary condition to implement efficient learning processes should not be restricted to any geographical or institutional entity.

5.6   References

ANCORI, B./BURETH, A./COHENDET, P. (2000): The economics of knowledge: is there a reasonable position between the absolutist position on codification and the absolutist position on tacitness?, *Industrial and Corporate Change*, forthcoming.

ARGYRIS, C./SCHÖN, D.A. (1978): *Organizational learning; a theory of action perspective.* Reading MA: Addison-Wesley Publishing Company.

BATESON, G. (1972): *Steps to an Ecology of Mind.* New York: Chandler Publishing Company.

BURETH, A. (1994): *Apprentissages et organisation: une analyse des phénomènes d'irréversibilisation.* PhD dissertation, Université Louis Pasteur, Strasbourg.

BURETH, A./WOLFF, S./ZANFEI, A. (1997): The two faces of learning by cooperating: The evolution and stability of inter-firm agreements in the European electronics industry, *Journal of Economic Behavior and Organization*, 32, pp. 519-537.

CHOO, C.W. (1996): The knowing organization: how organizations use information to construct meaning, create knowledge, and make decisions, *International Journal of Information Management*, 16, pp. 329-340.

GUILHON, B./GIANFALDONI, P. (1990): Chaînes de compétences et réseaux, *Revue d'Economie Industrielle*, 51, pp. 97-111.

HAHN, R./GAISER, A./HÉRAUD, J.-A./MULLER, E. (1995): Innovationstätigkeit und Unternehmensnetzwerke, *Zeitschrift für Betriebswirtschaft*, 65, pp. 247-266.

HÉRAUD, J.A./KERN, F. (1997): Les CORTECHS: Innovations, apprentissage en coopération et dynamique organisationnelle, GUILHON, B. ET AL. (Eds.) *Économie de la connaissance et organisation.* Paris: L'Harmattan, pp. 383-399.

HÉRAUD, J.A./MULLER, E. (1998): *The Impact of Universities and Research Institutions Labs on the Creation and Diffusion of Innovation-Relevant Knowledge: the Case of the Upper-Rhine Valley.* Paper presented at the 38th Congress of the European Regional Science Association, August 28-31 1998, Vienna.

HÉRAUD, J.A./NANOPOULOS, K. (1994a): *La sous-traitance industrielle en Alsace*, BETA, Université Louis Pasteur, Strasbourg, March.

HÉRAUD, J.A./NANOPOULOS, K. (1994b): Les réseaux d'innovation dans les PMI: illustration sur le cas de l'Alsace, *Revue Internationale PME*, 3-4, pp. 65-86.

KOFMAN, F./SENGE, P.M. (1993): Communities of commitment: the heart of earning organizations, *Organizational Dynamics*, 22, pp. 15-43.

KOSCHATZKY, K. (2000): A River is a River - Cross-border Networking between Baden and Alsace, *European Planning Studies*, 8, pp. 429-449.

KOSCHATZKY, K./HÉRAUD, J.A. (1996): Institutions of Technological Infrastructure, Final Report to Eurostat. Karlsruhe/Strasbourg: Fraunhofer ISI/BETA.

LUNDVALL, B.-Å. (1992): *National systems of innovation: toward a theory of innovation and interactive learning*. London: Frances Pinter

LUNDVALL, B.-Å. (1988): Innovation as an Interactive Process: from User-Producer Interaction to the National System of Innovation, DOSI G./FREEMAN, CH./NELSON, R./SILVERBERG, G./SOETE, C. (Eds.), *Technical Change and Economic Theory* , London: Pinter Publishers, pp. 45-67.

MARCH, J.G./SIMON, H.A. (1993): *Organizations*, 2nd ed. Oxford: Blackwell.

MILES, I./KASTRINOS, N./FLANAGAN, K./BILDEBEEK,R./DEN HERTOG, P./HUNTINK, W./BOUMAN, M. (1994): *Knowledge-Intensive Business Services: Their Roles as Users, Carriers and Sources of Innovation*. Manchester: PREST.

MULLER, E. (1999): *There is No Territorial Fatality! (or How Innovation Interactions between KIBS and SMEs May Modify the Development Patterns of Peripheral Regions)*. Paper presented at the 39th ERSA (European Regional Science Association) Congress, August 23-27, 1999, Dublin.

NONAKA, I. (1994): A dynamic theory of organizational knowledge creation, *Organisation Science*, 5, pp. 14-37.

SAVIOTTI, P.P. (1998): On the dynamic of appropriability, of tacit and of codified knowledge, *Research Policy*, 26, pp. 843-856.

SHANNON, C./WEAVER W. (1949): *The mathematical theory of communication*. Boston: University of Illinois.

STIGLITZ, J.E. (1987): Learning to learn, localized learning and technological progress, STONEMAN, P./DASGUPTA, P. (Eds.) *Economic policy and technological performance*. Cambridge: Cambridge University Press, pp. 125-153.

VON HIPPEL, E. (1988): Trading trade secrets, *Technology Review*, 2, pp. 58-64.

WEICK, K.E. (1979): *The social psychology of organizing*. New York: Random House.

6. Innovative Links between Industry and Research Institutes – How Important Are They for Firm Start Ups in the Metropolitan Regions of Barcelona, Vienna and Stockholm?

(Spain, Austria, Sweden)

Javier Revilla Diez

632 M13
L25

6.1 Introduction

Current discussions often describe young companies as innovative, expanding, and having a positive impact on employment. Their tendency to co-operate is strongly emphasised. Numerous regional-political measures on a local, national, and supra-national level attempt the inter-connection of local actors, while the main interest is focused especially on young companies and research institutes. Theoretical arguments furnish many concepts such as the knowledge based economy, the learning economy or the learning region, network approaches, spill-over approaches, etc. (Florida 1995; Mansfield/Lee 1996; Charles/Goddard 1997; Malmberg/Maskell 1995; Tödtling 1994). In simple terms, the argumentation is the following: due to growing competitive pressure, industry, and in particular young companies are forced to bring innovative products onto the market. The increasing complexity of the innovation process leads to increased co-operation with innovation-relevant actors such as customers, suppliers, service companies in close co-operation with industry, competitors, and research institutions. Since numerous founders have an academic background it is presumed that interwoven relationships with research institutes can easily be extended, and the innovation ability of the companies thus increased. Based on representative surveys, this contribution aims to show the real significance of research institutions for young companies. The observations are guided by the following four hypotheses derived from theoretical discussion:

1. Young companies stimulate structural change.

2. Young companies co-operate primarily with research institutions.

3. Research institutions give crucial impulses especially with the development of new products, and they are important partners to young companies in all stages of the innovation process.

4. Young companies co-operate mostly with local research institutions.

6.2 Data Base - The European Regional Innovation Survey (ERIS)

The basic data for the following analysis was collected in the framework of the DFG (German Research Association) focal programme " Technological change and regional development in Europe". In eleven European regions, relevant actors of a regional innovation system, i.e. companies from the processing industry, service companies working in close co-operation with industry, as well as research institutions, were questioned about their innovation and co-operation behaviour during a project jointly carried out by the universities of Hanover and Cologne, the Technical University of Freiberg, and the Fraunhofer Institute for Systems and Innovation Research. Questionnaires were used which were jointly formulated and targeted specifically at the different actors, which guaranteed a high level of comparability (Fritsch *et al.* 1998: 248). The analyses in this contribution are restricted to the three metropolitan agglomerations of Barcelona, Stockholm, and Vienna. The three regions were delimited according to identical criteria. Besides the central city, location of service companies working in close co-operation with industry, as well as research institutions, above all, the hinterland was also taken into account, which is generally more characterised by industry. In this way, it was possible to represent metropolitan innovation systems and thus to analyse a regional level which is gaining importance in the course of globalisation and the competition between regions which is linked to it. This contribution concentrates on the interpretation of the industrial data record. Following general questions about the plant, industrial innovation activities and the internal and external co-operation linked to it were considered. Questions about external co-operation were asked concerning the optional co-operation partners, i.e. customers, suppliers, service companies closely co-operating with industry, competitors and research institutions, following an identical pattern (as to whether co-operation takes place, and if yes, at which stage of the innovation process, how intensely, the co-operation partner's location). Young companies are defined as those which have been in existence since 1990; older companies were founded prior to 1990.

Table 6-1 shows the response rate of the sample surveys. According to Sachs' formula, all ranges of samples are representative. Particularly high response rates were reached in Vienna and in Stockholm, whereas the response rate in Barcelona was distinctly lower. In view of the fact that the Community Innovation Survey had to be broken off due to too low a response quota, the total number of about 400 replies must be seen as a success. In all three regions, research institutions showed the highest level of participation.

Table 6-1:   **Metropolitan innovation survey**

	Metropolitan Region of		
	Vienna	Barcelona	Stockholm
Surveys on:			
Manufacturing Firms			
Total Population	908	2,650	1,879
Responding Firms	204	394	451
Representativeness Ratio in %[a]	23	15	24
Producer Service Firms			
Total Population	648	598	1,301
Responding Firms	185	105	334
Representativeness Ratio in %[a]	29	18	26
Research Institutes			
Total Population	650	424	346
Responding Research Units	290	148	173
Representativeness Ratio in %[a]	45	35	50

Note **a**:   number of responding firms or research institutes divided by the total number of registered firms or research institutes multiplied by 100

A comparison between the three metropolitan agglomerations shows completely different start-up conditions (cf. Table 6-2). In Barcelona, the year the questioned industrial companies were founded is relatively equally distributed over the past decades. The questioned industrial companies in Vienna show a distinctly older age structure: about 60 % of the companies were founded prior to 1970. In contrast to Barcelona and Stockholm, no particular increase in company start-ups since 1990 is found. Stockholm has been showing increased start-up dynamics since 1990, after stagnant founding activities during the 1970s and 1980s.

Table 6-2:   **Year of foundation of the questioned industrial companies**

	Barcelona		Vienna		Stockholm	
	absolute number	%	absolute number	%	absolute number	%
Year of foundation:						
before 1970	102	27	114	57	153	35
1970 – 1979	84	22	33	17	67	15
1980 – 1989	97	25	24	12	94	21
since 1990	99	26	28	14	126	29
Total:	382	100	199	100	440	100

Source: ERIS 1997/1998

6.3   Empirical Analysis

First hypothesis: **Young companies promote regional structural change**

The importance of the effects which young companies or company start-ups have on structural change depends, among others, on the branch in which start-ups take place, and on the existing economic structure. If foundations help broaden and equalise the regional sectoral structure and thus the range of goods or services offered, then they have positive effects on the corporate stock of a regional economic system. Structural change is encouraged primarily through start-ups in growing and expanding economic sectors. This is particularly true for technology-oriented start-ups, which show a high degree of readiness for innovation and contribute to technological change. From a regional economic point of view, young technology-oriented companies are very important since they often concentrate on a specific region; they thus have an effect on employment and advance dynamic regional development (Kulicke 1993). Young companies are particularly important in former industrial regions, where they stimulate the impulses for re-structuring and re-orientation of the economy (e.g. Tichy 1981; Hamm/Wienert 1990).

Table 6-3:   **Composition of the interviewed plants according to industrial sector**

	Barcelona				Vienna				Stockholm			
	old[1]		young[2]		old[1]		young[2]		old[1]		young[2]	
	abs.	%	abs.	%	abs.	%	abs.	%	abs.	%	abs.	%
Sectors:												
Textile, clothing, leather	36	13	13	13	12	7	1	4	1	0	4	3
Food, beverages, tobacco	10	4	2	2	20	12	4	14	22	7	3	2
Wood, paper, printing	31	11	14	14	39	23	9	32	71	23	30	24
Chemicals, rubber, plastics	88	31	17	17	30	18	5	18	46	15	10	8
Electrical and Optical equipment	32	11	15	15	25	15	3	11	44	14	29	23
Basic metals and metal products	41	15	19	19	19	11	4	14	60	19	28	22
Machinery, transport, equipment	44	16	19	19	26	15	2	7	70	22	22	18
Total	282	100	99	100	171	100	28	100	314	100	126	100

1. Companies founded prior to 1990 are defined as "old firms"
2. Companies founded since 1990 are defined as "young firms".
Source: ERIS 1997/1998

At first glance, the results of the three metropolitan agglomerations cannot confirm the above-mentioned positive effects on structural change: a comparison between the composition of old and young companies according to branches reveals no significant difference between the groups. To a certain degree, path dependency is perceived, i.e. young companies come into existence in sectors which traditionally characterise the industrial structure of the specific region (cf. Table 6-3). Differences between old and young companies are slightly larger when looking at the technology fields they occupy. Although statistically insignificant, slight differences can be found regarding Vienna and Stockholm. In Vienna, the share of young companies in the field of bio-technology is a little higher than that of old companies, in Stockholm the same is true for the areas of micro-electronics, optical electronics and laser technology (cf. Table 6-4).

Table 6-4:   **Technology areas of the interviewed industrial companies**

	Barcelona				Vienna				Stockholm			
	old[1]		young[2]		old[1]		young[2]		old[1]		young[2]	
	abs.	$\%^3$	abs.	$\%^3$	abs.	$\%^3$	abs.	$\%^3$	abs.	$\%^3$	abs.	$\%^3$
Technology fields:												
Bio-technology, pharmacy	25	12	5	7	15	13	4	29	20	9	8	9
Chemistry	70	32	18	25	26	22	4	29	66	28	12	14
Energy technology	30	14	9	13	24	20		0	59	25	16	18
Information and communication technology	53	25	20	28	34	29	4	29	89	38	38	44
aeronautics and aerospace technology	1	0	1	1	2	2		0	7	3	2	2
medicinal and health technology	18	8	1	1	13	11		0	24	10	12	14
micro-electronics, optical electronics, laser technology	17	8	8	11	12	10	2	14	38	16	22	25
new material	103	48	33	46	62	53	7	50	128	55	47	54
production and processing technology	130	60	43	60	71	60	9	64	161	69	47	54
sensor technology, metrology, control engineering, analytic engineering	53	25	11	15	34	29	7	50	99	42	29	33
environmental technology	49	23	8	11	25	21	3	21	89	38	23	26
traffic and transport engineering logistics	30	14	12	17	32	27	5	36	60	26	21	24
cases	216		72		118		14		233		87	

1. Old companies are those founded prior to 1990
2. Young companies are those founded from 1990 onwards
3. indications in % of all cases
Source: ERIS 1997/1998

The results concerning general industrial ratios confirm previous studies. In all three agglomerations, young companies are distinctly smaller and show a higher average increase in employment figures than old companies. According to Nerlinger (1998), there is a close relationship between the development of employment figures and corporate innovative behaviour. He could prove that innovating companies reach faster growth of employment than non-innovating companies. This statement seems to be confirmed by the innovation indicators from the interviewed companies. The percentage of product-innovating companies is higher among young companies, which also distinguish themselves by a higher share of R&D expenditure. Moreover, young companies achieve a significantly higher share of their turnover with newly introduced products, which is not a surprising result, given the age of the companies (cf. Table 6-5).

Table 6-5:   Characteristics of the interviewed industrial companies

	Barcelona		Vienna		Stockholm	
	old[1]	young[2]	old[1]	young[2]	old[1]	young[2]
General industrial ratios						
Average company size (employees)	156	87	263	52	136	50
Average absolute development of the employment figure from 1994-1997	+1	+10	-15	+3	+1	+5
Share of High-tech companies in % of the total of companies[3]	39	33	43	39	48	47
Period of product life cycle (in months)	34	24	14	17	56	14
Share of co-operating companies in % of the total number of companies	75	88	64	57	67	64
Innovation relevant ratios						
R&D staff in % of total employees	7,9	5,8	16,2	7,4	8,2	12,2
R&D expenditure in % of the total turnover	4,7	5,5	3,7	5,9	4,4	6,3
Share of innovating companies in % of the industrial total (firms which spend more than 20 % of their total turnover on innovation expenditure)	27	41	22	36	19	22
Share of product-innovating companies in % of the industrial total	80	80	69	76	58	52
share of turnover held by newly introduced products (within the past three years)	22,3	36,2	17,2	21,5	23,6	36,1

1. "Old companies" are those founded prior to 1990
2. "Young companies" are those founded from 1990 onwards
3. High technology sectors as defined by Hatzichronoglou (1997)
Source: ERIS 1997/1998

Second hypothesis: **Young companies co-operate primarily with research institutions**

In view of the regional distribution of technology-oriented company start-ups, Nerlinger (1998) sees a positive correlation between public knowledge infrastructure and the number of technology-oriented start-ups. Here, the question arises as to what role co-operation between young companies and research institutions matters. Are research institutions the central contact available for the resolving of industrial innovation problems? A glance at the qualification of founders of technology-oriented companies shows that there are a great number of engineers and natural scientists with a university education among them (Nerlinger 1998; Schmude 1994). Such education can be supportive for the exchange of knowledge, as well as for co-operation relationships between founders and research institutions. Due to former studies or employment, personal contacts come into existence, which can be used for the establishment of co-operation relationships (Audretsch/Stephan 1996; Charles/Goddard 1997).

Table 6-6:   **Sources of information for the realisation of product innovations**

	Barcelona				Vienna				Stockholm			
	old[1]		young[2]		old[1]		young[2]		old[1]		young[2]	
	abs.	%[3]	abs.	%[3]	abs.	%[3]	abs.	%[3]	abs.	%[3]	abs.	%[3]
Source of information:												
Customers, buyers	200	83	72	83	125	87	24	100	219	92	85	90
Suppliers	109	45	38	44	71	49	14	58	123	52	64	67
Direct competitors	128	53	41	48	75	52	14	58	154	64	50	53
Research institutions	44	18	17	20	31	22	1	4	51	21	25	26
Companies Services	61	25	22	26	27	19	2	8	40	17	15	16
Fairs/Exhibitions	157	65	58	67	89	62	14	58	152	64	62	65
Technical literature	96	40	26	30	86	60	11	46	135	57	45	47
Media	47	20	14	16	33	23	7	29	52	22	20	21
Internet	25	10	1	1	5	4	1	4	30	13	21	22
Total	240		86		144		24		239		95	

1. Companies founded prior to 1990 are defined as "old firms"
2. Companies founded since 1990 are defined as "young firms".
3. Indications in % of all cases.
Source: ERIS 1997/1998

However, the empirical results qualify the above-mentioned possible co-operative links between young companies and research institutions (cf. Table 6-6). The results show that customers are by far the most important source of information for the realisation of product innovations. In addition, the interviewed companies named suppliers, direct competitors, specialised fairs, and technical literature (Revilla Diez 2000). An interesting fact is that the sources of information used by old and young companies strongly resemble each other. Even if research institutions are mentioned in the specialised literature, they are seldom integrated into concrete innovation projects. Regarding the actual co-operation relationships of the young companies interviewed, their customers are the most important co-operation partners (cf. Table 6-7). At a stage where both the clientele and the product base are limited, the companies concentrate on the customers' requirements. Young companies are much more oriented towards their customers than old companies are. Further partners are suppliers, as well as service companies working in close co-operation with industry. In all three regions, research institutions play a relatively insignificant role. In comparison with old companies, research institutions participate less in industrial innovation projects carried out by young companies. Only in Stockholm do research institutions represent an important co-operation partner for at least one quarter of the interviewed young companies.

Table 6-7:   **Co-operation links shown by the industrial companies interviewed**

	Barcelona				Vienna				Stockholm			
	old[1]		young[2]		old[1]		young[2]		old[1]		young[2]	
	abs.[3]	%[4]	abs.[3]	%[4]	abs.[3]	%[4]	abs.[3]	%[4]	abs.[3]	%[4]	abs.[3]	%[4]
Co-operation partners:												
Customers	141	66	67	77	52	48	13	81	153	73	68	84
Suppliers	115	54	55	63	39	36	7	44	78	37	29	36
Service companies	136	64	49	56	64	59	11	69	98	46	33	41
Competitors	42	20	26	30	26	24	4	25	41	19	13	16
Research institutions	55	26	15	17	27	25	2	13	79	37	21	26
Cases	213		87		109		16		211		81	

1. Companies founded prior to 1990 are defined as "old firms"
2. Companies founded since 1990 are defined as "young firms".
3. Number of companies co-operating intensely with their respective partners
4. Indications in % of all cases.
Source: ERIS 1997/1998

Third hypothesis: **Especially in the development of new products, crucial impulses come from research institutions, which, throughout the innovation process, are an important contact available to young companies**

Although research institutions are not among the preferred co-operation partners of the interviewed companies in the three metropolitan areas, the question is in which stages of the innovation process does co-operation take place. In contrast to the linear innovation model, i.e. the sequential process of research, development, production, and the marketing of new technology, the recursive interactive innovation model emphasises the crucial role of feedback loops between the linear model's different phases. Besides internal interaction in the innovating company, interaction takes place between companies, in particular between producers and users of new technology, as well as with the external system of science and technology (Kline/Rosenberg 1986). Consequently, according to the present theoretical discussion, innovations are no longer produced by individual companies but come into existence through a co-operation process between companies and their scientific and institutional environment. The results discussed up to the present could at least confirm the importance of co-operation for industrial innovation processes.

Table 6-8: **Co-operation relationships according to the different stages of an innovation process - Vienna**

	Indications in % of all cases									
	Customers		Suppliers		Service companies		Research institutions		Competitors	
	old[1]	young[2]	old[1]	young[2]	old[1]	young[2]	old[1]	young[2]	old[1]	young[2]
Stages of the Innovation process:										
General information exchange	45	81	34	44	47	63	20	13	21	25
Generation of new ideas	41	69	28	44	44	50	17	6	19	25
Conception/Pre-development	39	69	28	38	43	50	18	6	15	25
Prototype development	38	69	25	44	27	44	17	6	12	19
Pilot scheme	35	75	23	31	31	44	17	6	8	19
Market introduction	39	69	19	25	29	31	8	0	15	25

1. Companies founded prior to 1990 are defined as "old firms"
2. Companies founded since1990 are defined as "young firms".
Source: ERIS 1997/1998

Table 6-8 shows that all interviewed industrial companies in Vienna co-operate very closely with customers in all phases of the innovation process. Co-operation starts with an animated exchange of ideas at the beginning of an innovation project, goes through prototype and pilot scheme, and ends with market introduction. Compared with old companies, young companies show a distinctly higher tendency towards co-operation with customers in all stages of the process. Regarding research institutions, co-operation focuses on the early stages of the innovation process. A similar co-operation pattern is shown for industrial companies in Barcelona and in Stockholm (cf. Tables 6-9 and 6-10).

Table 6-9: **Co-operation relationships according to the different stages of an innovation process - Barcelona**

| | Indications in % of all cases | | | | | | | | | |
| | Customers | | Suppliers | | Service companies | | Research institutions | | Competitors | |
	old[1]	young[2]	old[1]	young[2]	old[1]	young[2]	old[1]	young[2]	old[1]	young[2]
Phases of the innovation process:										
General information exchange	59	70	48	56	52	47	19	11	15	23
Generation of new ideas	58	63	42	53	45	44	15	13	13	26
Conception/Pre-development	43	49	38	46	41	43	19	9	13	20
Prototype development	47	62	42	41	34	34	15	8	11	16
Pilot scheme	45	57	31	39	31	29	12	9	7	16
Market introduction	47	60	23	30	25	22	7	2	7	17

1. Companies founded prior to 1990 are defined as "old firms"
2. Companies founded since1990 are defined as "young firms".
Source: ERIS 1997/1998

Table 6-10: **Co-operation relationships according to the different stages of an innovation process - Stockholm**

	Data/Indications in % of all cases									
	Customers		Suppliers		Service companies		Research institutions		Competi-tors	
	old[1]	young[2]	old[1]	young[2]	old[1]	young[2]	old[1]	young[2]	old[1]	young[2]
Phases of the Innovation process:										
General information exchange	65	75	34	33	29	33	32	23	15	12
Generation of new ideas	61	78	31	31	33	31	30	21	16	12
Conception/Pre-development	56	64	27	26	33	28	30	17	17	10
Prototype development	58	67	27	25	26	22	19	14	15	12
Pilot scheme	49	52	23	17	18	20	16	9	10	7
Market introduction	49	46	18	16	22	16	9	5	9	10

1. Companies founded prior to 1990 are defined as "old firms"
2. Companies founded since1990 are defined as "young firms".
Source: ERIS 1997/1998

Fourth hypothesis: **Young companies co-operate primarily with local research institutions**

The connection of generally small and medium-sized companies with each other, with local education and research institutions and with supportive structures such as the public administration and chambers of industry and handicrafts in the same way as for regional networks, should lead to synergy effects which improve the competitiveness of the regional economy (Koschatzky 1995). In this way, for example, the existence of close intra-regional inter-linked relationships between innovation actors can have a positive influence on the duration of the development of new products or processes, and thus improve the efficiency of individual networking actors.

Innovations and innovative companies are the result of a collective, dynamic process which different actors from a region participate in, and which thus form a network of synergy-creating links (Sternberg 1998: 245). An innovative milieu results from interactions between companies, research and educational institutions, political decision-makers, institutions and employees. Here, informal contacts play an important role for the establishment of the necessary trust base. Close intra-regional co-operation enables for joint co-operative learning, which helps to reduce uncertainties and risks in the development of new products, processes, or organisational change. Spatial closeness encourages learning from and with each other, which takes place due to relationships with suppliers, the mobility of employees, and face-to-face contacts (Sternberg 1998: 246; Malmberg/Maskell 1995). Integration into regional networks is particularly beneficial to small companies, which face considerable problems in the realisation of innovations. Uncertainties are considerably reduced since information is collected and analysed in co-operation (Fritsch *et al.* 1998: 247; Schätzl 1996). In the light of this, it is expected that young founders, due to their academic background, aspire to co-operation with local research institutions. Many founders choose their company's location near to their former place of study or of work. Since most founders locate their company close to their residence, such a choice of location is considered probable where it is also the founder's place of residence (Schmude 1994; Sternberg 1988). This is explained by the founders' improved chances to be kept informed through their direct environment, as well as existing social inter-links.

No exceptional significance of the local level has been found in the analysed regions: the interviewed industrial companies co-operate with both local and national or international partners. Regarding the regional reach of co-operative links, only an insignificant difference is found between old and young companies. Just as the companies of Stockholm, those of Barcelona are significantly more oriented towards regional and national contacts than those of Vienna. The most international orientation is shown by companies from Vienna (cf. Table 6-11).

Table 6-11: **Reach of the interviewed companies' co-operative links**

Reach:	Barcelona				Vienna				Stockholm			
	old[1]		young[2]		old[1]		young[2]		old[1]		young[2]	
	abs.[3]	%[4]	abs.[3]	%[4]	abs.[3]	%[4]	abs.[3]	%[4]	abs.[3]	%[4]	abs.[3]	%[4]
Metropolitan region	202	27	82	25	94	30	15	29	162	25	59	26
Remaining country	278	37	126	39	76	24	12	23	287	44	104	47
EU	126	17	48	15	68	22	13	25	110	17	31	14
The rest of the world	152	20	70	21	76	24	12	23	95	15	29	13
Total	758	100	326	100	314	100	52	100	654	100	223	100

1. Companies founded prior to 1990 are defined as "old firms"
2. Companies founded since 1990 are defined as "young firms".
3. Number of companies which have intensive co-operation with at least one partner
4. Figures in % of the responses
Source: ERIS 1997/1998

Congruence is found between these results and the regional turnover distribution of the interviewed companies. Industrial companies from Barcelona and Stockholm are distinctly more oriented towards the local and national level than companies from Vienna. Both young and old companies in Vienna realise almost 50 % of their turnover with EU or world-wide customers (cf. Table 6-12).

Table 6-12: **Regional turnover distribution of the interviewed industrial companies**

	Barcelona		Vienna		Stockholm	
Data in % of the total turnover	old[1]	young[2]	old[1]	young[2]	old[1]	young[2]
Regional turnover distribution:						
Metropolitan region	30	39	44	40	23	33
Remaining country	47	46	28	39	49	44
EU	14	12	17	13	15	14
The rest of the world	10	3	11	7	12	8
Total	100	100	100	100	100	100

1. Companies founded prior to 1990 are defined as "old firms"
2. Companies founded since1990 are defined as "young firms".
Source: ERIS 1997/1998

Where co-operation with research institutions does occur, it is particularly the young companies in Barcelona which are oriented towards local research institutions. In Stockholm, young founders co-operate principally with research institutions on a national level, whereas young founders from Vienna co-operate with both research institutions on a local, national, and international level (cf. Table 6-13).

Table 6-13: **Reach of co-operative links between the interviewed industrial companies and research institutions**

	Barcelona				Vienna				Stockholm			
	old[1]		young[2]		old[1]		young[2]		old[1]		young[2]	
	abs.[3]	%[4]	abs.[3]	%[4]	abs.[3]	%[4]	abs.[3]	%[4]	abs.[3]	%[4]	abs.[3]	%[4]
Reach:												
Metropolitan region	46	35	13	50	20	30	2	33	52	33	16	35
Remaining country	43	33	6	23	15	23	1	17	73	46	18	39
EU	19	15	4	15	14	21	1	17	25	16	7	15
Remaining world-wide	22	17	3	12	17	26	2	33	10	6	5	11
Total	130	100	26	100	66	100	6	100	160	100	46	100

1. Companies founded prior to 1990 are defined as "old firms"
2. Companies founded since1990 are defined as "young firms".
3. Number of companies which have close co-operation relationships with research institutions
4. Data in % of total responses
Source: ERIS 1997/1998

6.4   Conclusion

This contribution is an analysis of the actual significance of research institutions for industrial innovation processes in young companies. Predominant questions are: to what degree are research institutions integrated into metropolitan innovation systems and, thus, what is their regional economic influence. Based on the survey of companies from the manufacturing industry, the following statements can be made about the above formulated hypotheses:

H 1: Young companies advance regional structural change

Founding dynamics in the analysed regions are very different. Whereas Barcelona and Stockholm registered many new foundations in the 1990s, founding activities in Vienna had come to a standstill. A comparison of branches and technology fields of the young companies reveals no significant difference with the traditional structure. At least the following two options of interpretation are possible: Firstly, founding dynamics show a certain path dependency. Most of the young companies are found in those branches which already characterised the industrial structure prior to their foundation. Secondly, successful regional structural change depends not only on newly founded companies. The three metropolitan areas of economic concentration in question are part of the most important economic areas of their countries and have recorded high growth dynamics during the past years. Consequently, existing companies in particular make an important contribution to regional structural change.

H 2: Young companies co-operate primarily with research institutions

The exceptional significance which literature often attributes to research institutions in view of industrial innovation processes, must be relativised. As was clearly shown by the survey results, vertical co-operation partners such as customers and suppliers, but also service companies working in close co-operation with industry, rank higher than research institutions regarding the support of industrial innovation processes. Only in Stockholm are research institutions made use of by companies to a noteworthy degree. This result is true for both young and old companies. Young companies excel through an even stronger orientation towards their customers than old companies.

H3: Especially in the development of new products, crucial impulses come from research institutions, which, throughout the innovation process, are an important contact available to young companies

Regarding industrial innovation processes in young companies, the primary significance of customers is also reflected by co-operation with them during the respective stages of innovation. Customers are the central contact available in all stages of the innovation process, from general information exchange to market introduction. Only in Stockholm are research institutions made use of to a noteworthy degree. The advantage of a strong orientation towards customers is that market signals can be promptly taken into account by young companies. However, regarding research institutions, the question arises as to what extent industrial innovation ability could be improved by more intensive co-operation with research institutions (for example by projects carried out jointly, advanced training), thus making life-long learning possible in the sense of a knowledge society.

H4: Young companies co-operate primarily with local research institutions

Co-operation between young companies and research institutions is almost identical to the co-operation pattern of old companies. In the case of co-operation with research institutions, only the local research institutions of Barcelona represent a central referral point. Co-operation partners of companies in Stockholm are predominantly national research institutions, whereas those of companies in Vienna are both local and international research companies. The reason for the strongly deviating results of Barcelona could be the lack of international companies there. In both the cases of Stockholm and especially Vienna, where the technological competitiveness is considered to be higher than that of Barcelona, it is shown that, regardless of their respective age, companies must make use of both local and global knowledge resources in order to preserve or extend their competitiveness.

6.5   References

AUDRETSCH, D./STEPHAN, P. (1996): Company-scientist locational links: The case of biotechnology, *American Economic Review*, 86, pp. 641-652.

BIRCH, D.L. (1979): *The Job Creation Process*. Cambridge, Mass.: Economic Research Division. PB 81-107062 (Microfiche).

CHARLES, D./GODDARD, J. (1997): Higher education and employment – linking universities with their regional industrial base, *Paper prepared for the Thematic seminar on Territorial Employment Pacts in Ostersund,* Sweden. 18.-19. September 1997.

FLORIDA, R. (1995): Towards the Learning Region, *Futures,* 27, pp. 527-536.

FRITSCH, M./KOSCHATZKY, K./SCHÄTZL, L./STERNBERG, R. (1998): Regionale Innovationspotentiale und innovative Netzwerke, *Raumforschung und Raumordnung,* 56, pp. 288-298.

FRITSCH, M./SCHWIRTEN, C. (1998): Öffentliche Forschungseinrichtungen im regionalen Innovationssystem, *Raumforschung und Raumordnung,* 56, pp. 288-298.

HAMM, R./WIENERT, H. (1990): *Strukturelle Anpassung altindustrieller Regionen im internationalen Vergleich*. Schriftenreihe des Rheinisch-Westfälischen Instituts für Wirtschaftsforschung. Essen, Berlin.

KLINE, S.J./ROSENBERG, N. (1986): An Overview of Innovation, LANDAU, R./ROSENBERG, N. (Eds.) *The Positive Sum Strategy: Harnessing Technology for Economic Growth*. Washington DC: National Academy Press, pp. 275-305.

KOSCHATZKY, K. (1995): Utilization of innovation resources for regional development – empirical evidence and political conclusions, NATIONAL INSTITUTE OF SCIENCE AND TECHNOLOGY, SCIENCE AND TECHNOLOGY AGENCY (Ed.): *Regional Management of Science and Technology*. Tokyo: STA.

KULICKE, M. (1993): *Chancen und Risiken junger Technologieunternehmen: Ergebnisse des Modellversuchs "Förderung technologieorientierter Unternehmensgründungen"*. Heidelberg: Physica-Verlag.

MALMBERG, A./MASKELL, P (1995): *Localised learning and industrial competitiveness*. Paper presented at the Regional Science Association European Conference on "Regional Futures: Past and Present, East and West". May 1996.

MANSFIELD, E./LEE, J.-Y.(1996): The modern university: contributor to industrial innovation and recipient of industrial R&D support, *Research Policy,* 25, pp. 1047-1058.

NERLINGER, E. (1998): *Standorte und Entwicklung junger innovativer Unternehmen. Empirische Ergebnisse für West-Deutschland*. ZEW-Wirtschaftsanalysen. Schriftenreihe des ZEW. Band 27. Baden-Baden: Nomos Verlagsgesellschaft.

REVILLA DIEZ, J. (2000): The importance of public research institutions in innovative networks - empirical results from the metropolitan innovation systems Barcelona, Stockholm and Vienna, *European Planning Studies* (accepted and forthcoming in 2000 No. 4).

SCHÄTZL, L. (1996): *Wirtschaftsgeographie I. Theorie*. 6. Auflage. Paderborn, München, Wien, Zürich: UTB.

SCHMUDE, J. (1994): Qualifikation und Unternehmensgründung. Eine empirische Untersuchung über die Qualifikationsstrukturen geförderter Unternehmensgründer in Baden-Württemberg, *Geographische Zeitschrift*, 82, pp. 166-179.

STERNBERG, R. (1988): *Technologie- und Gründerzentren als Instrument kommunaler Wirtschaftsförderung*. Dortmund: Dortmunder Vertrieb für Bau- und Planungsliteratur.

TICHY, G. (1981): Das Altern von Industrieregionen – Unabwendbares Schicksal oder Herausforderung für die Wirtschaftspolitik? *Berichte zur Raumforschung und Raumordnung*, 31, pp. 3ff.

TÖDTLING, F. (1994): The Uneven Landscape of Innovation Poles: Local Embeddedness and Global Networks, AMIN, A./THRIFT, N. (Eds.): *Globalization, Institutions and Regional Development in Europe*. Oxford: Oxford University Press., pp. 68-90.

Section III: Innovation Networks in Transition

7. Implementation of a Network Based Innovation Policy in Central and Eastern European Countries – Slovenia as an Example

Günter H. Walter

7.1 Introduction[1]

Since the beginning of the 1990s, most Central and Eastern European Countries (CEECs) have achieved considerable success in the transformation to an open and innovation-oriented market economy. Nevertheless, the transition period is not over. As governments are still under the threatening trade-off between stabilisation policy and structural transition, salient issues of the new economic order such as privatisation, reform of the social security system and deregulation are not pursued consistently. Furthermore, there is the need for a long-term strategy for technological development and economic growth. In this situation many initiatives of international co-operation aim at supporting these countries through Western expertise. One example is the bilateral techno-scientific co-operation between Germany and CEECs (Abel 1999).

With the cessation of the previous socialist order, the CEECs were directly confronted with the task of fundamentally transforming their political and societal systems. This systemic transformation is historically unique in its radicality. Against a background of political instability, a positive development of the economy gains particular importance. The economic systems of these countries are now re-orienting themselves, away from a more or less centrally planned economy and from embedding in the socialist state system towards a free market economy; they are opening up to the global economy. At present, most of them are only able to gain a modest foothold in the world market.

In the time immediately following the changeover, national and international initiatives were - and still are today - simultaneously faced with a multiplicity of tasks in conditions of extreme shortage of resources. No clear strategies or policy recom-

---

1 This paper is partly based on (Walter/Bross 1997) but updated and modified.

mendations could be derived from existing political and economic models. However, in view of the international competition of locations for scarce resources, an overall political concept is needed as a prerequisite for successful economic development. An integrated plan to re-structure the whole economy is more relevant than the use of individual policy instruments. Three development courses will shape the future order: firstly, parts of the old system will survive, with their actors and institutions; secondly, economic policy models will become adopted which are mainly Western in character and, thirdly, endogenous potential will be enabled to develop in a way which was not possible under the imposed former order. The transformation of an economy requires action at very different policy levels. One important starting point for a country's innovative power and its economic success in the long term is innovation and technology policy.

Technology policy and innovation policy include all public measures which are oriented towards converting technical inventions into industrial applications and which support the diffusion of product and process innovations (Meyer-Krahmer/Kuntze 1992). Instruments of public technology policy include measures such as the institutional support of research institutions, financial incentives for industrial innovation projects, and the initiation and expansion of an innovation infrastructure in terms of consulting, technology transfer and innovation financing.

This paper attempts to select from the accumulated knowledge concepts relating to success determinants in innovation and technology policy that can appropriately be adapted for use in the process of modernisation now taking place in CEECs. Technology and innovation policies in these countries clearly have to pursue strategies of modernisation which support the internationally opening up processes of these countries which mobilise their endogenous resources and form the basis for an integrated policy concept. Western countries can only provide very limited assistance for this transition process, in the form of "help towards self-help". But even this approach always has to take account of the differing initial situations in the individual CEECs and the differing paths they have adopted in the process of reform.

7.2   Strategy: Bottom-Up, Endogenous Growth

By opening up their economic systems, CEECs have joined international competition. After several decades of political and economic encapsulation from the Western world, consensus has been reached with regard to the necessity for their integration into the global economy and into the international division of labour. This integration can best be realised by a "free market" type of economic system which derives its impetus from individual, decentral initiatives. The realisation of existing competitive potentials is hampered by the current state of the markets, which are not yet fully able to function, and by the existence of certain types of market failure.

This situation also requires a comprehensive public technology and innovation policy. However, before discussing the instruments that should be used in innovation and technology policy, and the extent to which the government should intervene, it is necessary to consider which modernisation strategies should be adopted in view of national framework conditions.

Integration into the international division of labour can occur in different ways, as various economic theories suggest. For CEECs it is possible to imitate leading industrialised nations or approach their standards of living by producing innovative products that are mature for the market. To ensure a competitive industry, most advanced market economies have developed and built up a national innovation system consisting of dynamic business organisations (e.g. innovative industrial firms of all sizes and sectors, new technology based companies, innovation supporting services in consultancy and financing), a science, research and educational sector (research and development, academic education, vocational training) and a differentiated framework of research, technology and innovation policy instruments to provide a wide range of research and development (R&D) and to support links between R&D and industry.

Empirical and theoretical research offers starting-points for the derivation of strategies and policy recommendations to build up a national innovation system. One basic element here is the establishment of innovation networks. Economic theories suggest that the development of modern technologies is characterised by a growing interdependence and complementarity of different areas in society such as R&D and industry. Studies confirm the relevance of these factors for the economic success of a location. In regions which are economically highly developed, networking between the different actors is very strong (Herden 1992).

Management research has also thoroughly investigated the importance of strategic network relations, particularly between different steps of the value-added chain (Sydow 1992). Networks serve the purpose of interlinking actors in production, services and research in such a way that their comparative strengths are exploited to the full and developed further. Innovation networks are able to activate, co-ordinate and combine the resources which support the technological competitiveness of regions and countries.

Economic theory also provides a valid explanation for this phenomenon in the network theory. Generally speaking, in these models complementary - and therefore resource-saving - learning processes are initiated between the actors in the economic process. Firstly, "learning by doing" and "learning by using" take place between suppliers, producers and customers in their business relations (Kline/Rosenberg 1986). But networks are not only characterised by performance-related businesses. In the course of repeated interchanges the integrity of the partners is recognised; a relationship of trust is built up, and stable personal relationships also develop between the partners (Walter 1992). Through the co-operation with the network partners, knowledge is accumulated which forms the basis for future competitive advantages. Externalities in an alliance of this kind extend beyond the reduction of transaction costs: dynamic economies of scale in terms of learning and complementary investments enhance the productivity achieved with limited resources. Moreover, the reduction of uncertainties through institutionalisation of the exchange relationships, and risk share among several partners in case of failure, are very important aspects, particularly in industrial innovation activities. These are the factors that make networks successful and have caused them to be an object of economic and regionally-oriented research for several years now.

The results of various studies indicate that the parameters of firm size, industrial sector and technology orientation give rise to different patterns and intensities of co-operation (Koschatzky 1998). Innovation networks are based on specific national, regional or local development patterns in the sense of different "best practices" in technogenesis and technology use; the starting-point for these is formed by specific innovation potentials, such as accumulated knowledge in certain technologies. A synergetic innovation network arises e.g. through the alliance of actors in research, production and services, aiming at the optimal exploitation of existing resources for growth (Koschatzky 1999). Overlapping, flexible network relations have proved to be effective in coping with structural changes caused by shifts in the framework conditions for competition (Herden 1992).

From the viewpoint of innovation economics, national and regional networks can be used for the economic development of CEECs: to systematically exploit existing development potentials by converting them into application-oriented knowledge and the rapid diffusion of new technologies (Koschatzky 1997), thus also providing a good basis for participation in international networks. The innovation networks that are of particular interest for innovation and technology policy arise through co-operation between suppliers and users of technological knowledge: these may for example be relations between enterprises and technology suppliers (e.g. universities and research establishments). Additionally such networks also refer strongly to the exchange of business know-how and innovation financing by public R&D promotion programmes or private venture capital.

The network concept can provide starting-points for the application of a specifically-oriented innovation and technology policy: firstly, the mobilisation and complementation of resources for the development and application of new technologies; secondly, the co-ordination and interlinking of these resources within innovation networks involving all the relevant actors in industry, science and policy; thirdly, the integration of networks into the national and international development and production of technology, by the creation of active interfaces and the support of co-operation.

Innovation and technology policy in particular is able to support innovative co-operation between science and industry, if it succeeds in integrating the relevant actors into a network of enterprises, research institutes, universities and innovation services. Having a sound base in science and also in industry, technological and economic modernisation of CEECs can rely mainly on endogenous potential within the countries. Growth in firm size, employment, export opportunities can be realised. By choosing a strategy that supports innovative firms significant positive economic and social effects can be expected, especially if all relevant fields of policy are integrated into an innovation-based overall industrial restructuring.

7.3   Creation of Networks in Central and Eastern Europe

This chapter commences with a brief description of societal framework conditions in CEECs. Following and based on this, networks are then specifically described and an example is given of ways to implement them in Central and Eastern Europe.

7.3.1   Shortcomings of Innovation Systems in Transition

Under the socialist economic order, research and industry in the CEECs were char-acterised by a centrally-steered, "top-down" type of organisation which was pre-dominantly state-controlled. This situation led to vertical structures in science and industry, with very few relations of horizontal interchange.

Exchange within networks must occur as interaction between the various actors in the national economy. Thus what is required, rather than the former centralistic policies, is a policy strategy which emphasises the free development of individual actors according to the "bottom-up" principle. In the foreground is the promotion of efficient, innovative co-operation in the form of "horizontal" relations between ac-tors especially in industry and research. In strategic terms, this implies that policy should aim to support interactive relations with a view to the formation of networks, and should itself become active in the creation of new networks.

In CEECs, innovation and technology policy measures that are intended to support networking activities are sometimes still influenced by the surviving remains of inherited political and legislative framework conditions. Another problem arose in the first few years of the transitional phase when, due to the shortage of resources, drastic cutbacks in funding took place, with the result that today much of the previ-ously-existing potential in science and industry is on the point of collapse. Strategic plans or financial means are only available to a certain extent for their consolidation during the system changeover.

Industry and science

Industrial structure in CEECs, with its large-scale business units, is disproportionate to the small domestic markets. Capital-intensive production was previously based principally on mass production and economies of scale; only some products offered satisfied international standards. In all CEECs, the situation is marked by drastic declines in production in the first few years of transition, due to the disappearance of the trade relations formed under the socialist regime. However, today one sometimes already finds a wider variety of small businesses. The financial means at their disposal are small. They concentrate on goods and commodities for everyday use. Small-scale production is based on forms of production which still represent "manufacture", i.e. technical crafts; in these countries, manual skills and corresponding capabilities predominate.

For CEECs, pressure to adapt to market forces is a new challenge, further intensified by foreign competition. In some cases the managers of firms have been able to make autonomous, market-oriented decisions and develop individual strategies for survival. In view of the high degree of uncertainty regarding markets and the shortage of resources, such strategies tend to be based on improvisation skills and lead to relatively simple, not very technology-intensive production structures and to small production volumes (Portratz/Widmaier 1999).

In industrialised countries in the West, small and medium-sized firms which are innovation-oriented or capable of innovation play an important role in economic development; in most CEECs, however, such firms are not yet numerous, as an after-effect of ownership law under former socialistic regimes and a previous lack of societal acceptance. Therefore, the re-structuring of the industrial sector is still incomplete in CEECs. Whereas on the one hand the start-up and survival chances of firms are still very hazardous, suppliers at the various stages of the value-added chain also appear underdeveloped, as does industrially-oriented research.

In general terms, until now this has meant that the private sector was virtually unable to fulfil tasks in the field of R&D to a larger extent. Moreover, industrial R&D in CEECs was previously mainly concerned with carrying out adaptation developments, and was not oriented towards innovative products or new production technologies. Industrial R&D potential has turned out to be one-sided in its qualification. In addition, it has suffered to some extent from the fragmentation of industrial complexes and from cuts in personnel. For firms whose survival strategies do not lie in the area of sophisticated technologies, future prospects are at best uncertain.

Technological R&D potential does exist, however, in the public research institutions. In CEECs the performers of research were primarily the universities, public research institutes and academies. Due to the intensive research that took place in publicly funded institutions, there was an ample availability of R&D results. Today, many CEECs still possess a broad range of research institutions. Some of these countries cultivate a presence in various different areas of basic research. For them, orientation towards the international scientific community both was, and is, a priority consideration. The universities and other scientific institutions regard themselves as an academic elite and, consequently, do not see themselves as "pre-thinkers" - or even problem-solvers - for industry, especially as private enterprise is often not considered as a potential partner or client of the science sector.

Not only does the vertical structure of the research landscape separate industry from science; often its effects are also felt within the science sector itself. All in all, the exploitation of research potential linked with industrial know-how in application-oriented research is too low.

Socio-political framework conditions

Policy and administrative law in CEECs does not provide incentives for innovation. Initiative and a willingness to bear risks and participate in free market competition are not sufficiently recompensed in terms of economic success. There is frequently a lack of generally valid regulations, particularly in the area of contract law, and a lack of (administrative) provisions for the legal enforcement of contracts in cases of conflict. The uncertainties with regard to these legal aspects constitute an obstacle to formal co-operation relations between actors, and negatively affect the subjective perception of success prospects for innovation projects in the private sector.

Often, policy regulations hardly allow for free communication or the free combination of resources. Policy is often still centralistic, fairly inflexible and not very demand-oriented. Practically-oriented politico-administrative decisions are frequently impeded by a rigid adherence to the "letter of the law" in the implementation of regulations, and by time consuming decision procedures. In CEECs, unlike countries in the West, it is not regarded as the self-evident duty of the Government to make (scientific) knowledge available to the general public or industry. This situation is rendered all the more serious by the fact that in these countries the state-owned institutions would be best positioned to initiate co-operative synergies between societal groups such as industrial enterprises and the science sector. Although in many CEECs the establishment of a new political order, including administrative and economic policy regulations has still not been completed, some CEECs have to a large extent resolved many of the problems associated with transition.

It is in this complex, multi-faceted context that measures for an operative and strategic innovation and technology policy have to be developed for individual CEECs. This has to be accomplished in a situation of extreme shortage of resources and in the face of other urgent and pressing policy requirements (e.g. structural assistance for regions in need, payments to the unemployed). Thus for most of these countries, it would be generally true to state that since the beginning of the transition, an innovation or technology policy has existed only in a rudimentary form, if at all, and that existing innovation potentials are endangered.

CEECs should build up innovation-supportive relations between all relevant contributors of resources in society. These relations include the formal and personnel exchange of information, networking and co-operation. The "mental gap" between science and industry must be eliminated in order to effectively exploit endogenous potentials. There must be greater awareness of the necessity to orient research more strongly towards the needs of industry. Up to now operational concepts have been lacking and co-operation have failed due to financial bottlenecks of the enterprises. The utilisation of technological research results for the development of innovative products necessitates co-operation between science and industry, with relations taking the form of an intensive two-way exchange in which users of technological knowledge test out its suitability for industrial manufacturing, and the necessary modifications are made in a process of mutual learning.

7.3.2 The Implementation of Networks in CEECs

The purpose of a network is to support industrial innovation by making available necessary resources such as technological and economic know-how, demand oriented funding for the promotion of R&D, production and market introduction of products and processes based on new technologies. To do so the network has to link all relevant actors: enterprises, institutions for technological R&D, for techno-scientific information and further qualification, and technology consulting. Also, entities for innovation financing (in both the public and private sectors), for innovation management and market consulting (including market research) need to be network partners. The network has to be extensive enough to provide support for enterprises all over the country as companies in all areas may need innovation services. The network must be oriented towards an internal exchange of funds, information and services, i. e. towards co-operation between all participants. Exchange of experiences is important for orienting services towards the real demand; it also implies the possibility of reversing the role of users and suppliers of technologies, services etc.

The successful implementation of a network concept in CEECs will not necessarily result from the transfer of measures that have proved successful in other countries. The same activities may have very different impacts when applied under different specific societal and political framework conditions. However, it is possible to identify success factors that are independent of any specific system and adapt them to different societal conditions (Walter 1992). This should be borne in mind when transferring experiences to CEECs and in the implementation of transition assistance by Western countries. Networks have arisen in Germany and other Western industrialised nations over a relatively long time span, and mostly through trial and error. For reasons of time and economy of resources, a trial and error process is not suitable for CEECs. Furthermore, due to the framework conditions described above, it cannot be assumed that in CEECs innovation networking between the various different sources of innovation potential will automatically occur.

Based on empirical experience, a possible procedure is now described for the infrastructurally-supported initiation and strengthening of national networks, with the possibility of integrating them into international innovation networks. This procedure takes account of already existing institutional starting points and relevant personal and political contacts in the countries concerned, but also supports early self-organisation of the networks.

Government innovation and technology policy can support efficient, innovative co-operation in the form of "horizontal" networks between research and industry, if it succeeds in integrating the relevant actors into the networks: enterprises, research institutions, universities and suppliers of innovation services. This can be done by strengthening existing interactive relations and initiating new networks on the one hand, and on the other by identifying network deficits. If such deficits are found, the missing network partners can be established by state initiatives as a part of innovation and technology policy.

The success or failure of innovation and technology policy measures supporting the network concept is decisively dependent on reaching a broad consensus of all relevant actors in policy, industry and science at an early stage (Koschatzky 1997). It is also important to jointly identify priority problem areas and fields of action. Concrete policy measures for support should be defined on this basis as well.

Networks support industrial innovation if they enable a demand-oriented exchange of techno-economic know-how to take place and mobilise funds. First, suppliers of know-how can be networked with one another and with know-how users. The suppliers of know-how are primarily application-oriented research and development establishments, techno-economic institutions and higher education institutions, but also - insofar as they (still) exist - development departments and research groups in industry. The main users are enterprises of various sizes in different sectors. For activities to reflect real needs, there is a necessity for close co-operation between suppliers and users and for interactive supplier-user learning processes with alternation of roles. For the mutual exchange of information, services and funds to take place, spatial proximity of the actors is also important.

Under the conditions that pertain in transitional countries, the responsibility for initiating and stimulating networks tends to lie with governmental agencies. These should entrust specific tasks to network actors according to their specialist expertise, their capacities and their location or radius of action, and should partly finance these tasks in the initial phase. This is the point of application for innovation and technology policy instruments designed to support the expansion and formation of the network and promote co-operation activities between the network partners. As well as the institutional promotion of important institutions in the network, financial incentives will result in learning effects and will spur on other initiatives - including private ones.

A strategy for the formation of an innovation network must involve all relevant actors at various regional and national levels and in different industrial sectors. Although the networking relations that arise are between decentrally active participants, it does appear important to have a central institution in the initiation phase.

Such an institution can perform planning and co-ordination tasks in the network and can provide organisational support. However, this institution should not function as a centralistic planning body - rather, its importance should be in acting as a moderator in the generation of a modernisation strategy and the formation of consensus among all relevant actors in science, industry and policy. Network co-ordination requires techno-economic competence and an abundance of contacts with users and suppliers of innovation support services, in order to collect information and the identify demand for them. An institution, as the nodal point of the network, also acts as a contact partner for all other network partners and establishes active external contacts, for instance to international networks. An interface of this kind gives the network access to the direct use of globally available research results and, conversely, enables it to co-operate on equal terms in the international exchange of knowledge and know-how by making its own resources available. For CEECs, this would seem an important contribution towards integration into global networks and gaining a position in the international technology competition.

The decentral elements of the network structure to be established include public teaching and research institutions as well as industrial and sectoral associations. These can make their sectoral or specific knowledge and know-how available. Transfer and advisory offices can cover different specialist areas and contribute at a national level to a comprehensive, complementary offer of knowledge and know-how. The wealth of highly specialised information contained in these institutions should be used by all network partners. As well as the technological input, surviving links and contacts to international science that may still exist in research are also important for the network.

Regional contact offices which are spatially accessible have to be available or be set up to provide users in the region with demand-oriented information and funds and to mediate contacts. These offices should be run by existing institutions (e.g. economic promotion agencies) which, as actors at a regional level, have the advantage of intensive awareness and are well suited for organising the exchange of specialist information. They should also be in a position to smooth out, at an informal personal level, possible differences that arise between network partners.

It is a good idea for the internal flow of communication and financial means within the network to be secured and organised by the co-ordinating institution. Also additional services should be provided to foster the exchange of experiences and the mediation of contacts. It seems important that communication is not "centralistic", but that all the actors intercommunicate. The contact offices also function as intermediaries, i.e. between the firms and suppliers of know-how. If the need arises, the network brings in other additional institutions. Existing gaps in the network are closed hopefully by policy support. Care should be taken to ensure their practical orientation, so that their services are accessible to all partners.

To sum up: Since innovation and technology policy in CEECs is only able to implement measures involving relatively low financial resources, these measures should be directed towards the initiation of networks. Networks should aim at mobilising and focusing existing institutional and personal resources in order to strengthen industrial innovative activity, and to stimulate firms which are as yet non-innovating, to engage in innovative activities. Financial resources can be used for the promotion of specific co-operation between partners in the network and institutional funding to close gaps in networks. In this respect, public financial assistance should be regarded primarily as "initiation financing".

7.4 FhG-ISI Scheme of Transfer of Institutional Know-How

The bringing together of existing resources and the initiation of innovative networks is a difficult task under present conditions in CEECs. Thus it appears important for developed industrialised countries to offer Central and Eastern Europe assistance in the process of transition and give "help towards self-help" to public organisations there. Such kinds of assistance can support the planning and implementation of adequate research, technology and innovation policy by providing analyses and new methods, and by the transfer of expert knowledge, training and advice. A scheme of this kind of policy consultation was developed by the Fraunhofer Institute for Systems and Innovation Research (FhG-ISI) at the request of the Federal Ministry for Education and Research. The scheme is based on empirical and theoretical know-how and can be flexibly adapted to the transitional context of individual CEECs. Transitional support of FhG-ISI for CEECs aims to stimulate modernisation processes based on the existing strengths of these countries. The governments in CEECs themselves have the responsibility for the individual steps and for their co-ordination.

Assistance by FhG-ISI in the implementation of a network based technology and innovation policy usually begins with an analysis of existing information in the form of a compact descriptive profile. Discussions with actors from policy, industry and science of the country concerned aim at further steps of co-operation. This exchange of experiences also serves as an opportunity for a transfer of basic information about modern western technology and innovation policy to CEECs. Personal contacts help to form a broad consensus on proposals for improvements. A suitable overall concept on policy measures to initiate and expand a network for the support of industrial innovation is elaborated by CEECs representatives and FhG-ISI in a joint development. Another part of the transition assistance requires binding commitment by the CEECs to building up networks: New institutions have to be created and topics covered by existing institutions have to be expanded or redefined.

In general, transitional assistance for CEECs is characterised by numerous, parallel tasks with different time horizons, fluctuating determinant parameters and a changing of contact partners and situations. This constellation overlays a basic structure with a multitude of personal dependencies, resulting in low flexibility and mobility. Thus, on the one hand there is a necessity for a long-term, integrated approach in transitional assistance, with gradual realisation in successive steps and the possibility of correction; on the other hand, there is also a need for relationships of interchange, the use of changing procedures and powers of improvisation.

7.5 German Transitional Assistance for Slovenia

As an example, this chapter describes German transitional assistance given to Slovenia. Since independence in 1991, Slovenia has established democratic institutions and achieved economic stability. Historically, attitudes were biased towards science, and interest in innovation only existed to a minor extent. Therefore, there is a lack of networking and co-ordination between the actors relevant to innovation.

7.5.1 FhG-ISI Transfer of Institutional Know-How to Slovenia

The support for Slovenia (Walter 1995) is a "typical" science and empirical-oriented technology and innovation policy advisory project by FhG-ISI. Its basic outline also applies to the work of the FhG-ISI in other countries and regions in Central and Eastern Europe.

In Slovenia, first activities took place in 1993 with a preparatory evaluation of the Slovenian situation based on information existing in Germany. Data about Slovenia relating to policy, economy, science and spatial structure were collected and interpreted. This analysis later served as a basis for the extensive "inventory" of the initial situation of the country. An introductory workshop was held in Slovenia, in which possible work steps and parts of the network approach were presented by FhG-ISI to the main actors in Slovenian industry, policy and science. In addition, the German and European innovation systems and especially the "promotion landscape" in the areas of technology and innovation policy were sketched. The next step in Slovenian-German co-operation was a screening of industry and science in Slovenia to identify areas and potentials which were to be integrated into an innovation network system of Slovenia. This assessment revealed the following picture of the situation: Despite good overall economic development, innovative networks needed to be further developed. Also special efforts had to be made to ensure that a modernisation concept for Slovenia was accepted and supported by all relevant social groups in economy, research, and policy.

FhG-ISI provided support to the Slovenian Ministry of Science and Technology to extend the previous considerations and activities in terms of R&D, innovation and build-up or enlargement of innovation-oriented networks. Thus, the Slovenian capabilities in technology and innovation policy could be improved by new methods of analysis, promotion steering and evaluation of projects, instruments and programmes. FhG-ISI also offered consultation and training on subjects such as technology foresight and new evaluation methods. To support co-operation between research, innovation funding and economic promotion, research institutes and other entities were trained e.g. in innovation management, utilisation of R&D results or setting up technology transfer and technology advisory groups.

As governmental agencies gained importance (Walter 1999), in 1997 and 1998 FhG-ISI efforts were integrated to the Slovenian Innovation Agency (SIA) project financed by the European Union (EU). FhG-ISI was involved in the conception and start up phase of this agency. The EU project was not only concentrated on counselling and advising but also provided financial means for institutional funding of SIA to act as a network co-ordination unit (Walter *et al.* 1997).

7.5.2   The Slovenian Innovation Agency

Network based policy in Slovenia focussed primarily on improving the already available innovative structures, public R&D programmes and potentials in economy and science. A specific network co-ordination unit - the SIA –should also create a more positive attitude in Slovenian society towards the necessity of industrial innovation.

Participants to the network managed by SIA should be

- Slovenian ministries that are responsible for science and technology and innovation policy (Ministries for Science and Technology and for Economic Affairs), e.g. to support co-operation between industry and science by providing financial help for joint R&D and contract research;

- technology transfer centres etc. as entities to mediate the demand for R&D of technology utilisers such as industrial enterprises and the supply side of R&D such as universities, R&D institutes;

- general business services and regional network institutions all over the country, such as existing economic development organisations, business services, industrial organisations (e.g. chambers of economy and crafts, economic promotion agencies).

Additionally, SIA should be involved in administering and managing the funding of public R&D programmes.

SIA was established in stages and started as a co-ordinating unit for public bodies that support industrial innovation and fulfil administrative tasks of funding programmes, beginning with the support of the operational management of a small subsidy programme on behalf of the Ministry of Science and Technology.

SIA started working in autumn 1997, as a unit within the Subdivision for Technology and Innovation of the Slovenian Ministry of Science and Technology. SIA received staff training in Slovenia and in EU countries and the staff of the SIA visited companies, ministries and other possible network actors. SIA also organised events in relation to EU access of Slovenia and elaborated a business plan for its future work.

Later in 1998, SIA activities stopped or were partly integrated into the usual administrative work of the ministry. Reasons were legal problems: the law of science and technology was delayed in parliamentary discussion. Today, SIA is no longer acting in network co-ordination. However, a continuation of the activities can be expected in the near future, based on a new law on science and technology in Slovenia as regulations on independent agencies also are a part of this law (Kalin 1999).

7.6   Looking Ahead

Central and Eastern Europe can only achieve international competitiveness if an innovative national economy is present. Taking account of the institutional situation and the specific strengths of CEECs, the network approach demonstrates concrete possibilities for making an effective contribution to the economic development and helping to build up international competitiveness. Network theory offers concrete starting points for the promotion of co-operative development in these countries. The targets of a network based technology and innovation policy in CEECs are: activation, focusing and complementation of existing potentials. In Slovenia and some countries, the first steps towards the implementation of networks have already been taken.

The co-operation between CEECs and Germany permitted the setting up of a partnership through which institutional know-how from advanced market economies could be transferred to these countries. During the course of co-operation, the CEECs developed an awareness of the requirements of modern technologies, and the bottlenecks of technology and innovation policy were perceived.

Despite the positive perspective on building partnerships and mutual learning between CEECs and Western countries, international collaboration with CEECs has to acknowledge that there are factors which cannot be thought of right at the beginning of the co-operation, but which may deeply affect its outcome. These factors in CEECs are especially the fragmentation of relevant actors, formal structures being very influenced by alternative streams which interfere with policy formulation and implementation, and furthermore, the high nontransparency of informal structures which still have to be overcome.

7.7 References

ABEL, E. (1999): *Bilateral Co-operation between Germany and CEECs: An Example of Sustainability*; Mimeo. FORUM BLED '99, Bled, 6–9 June 1999.

BROSS, U. (2000): *Innovationsnetzwerke in Transformationsländern*. Heidelberg: Physica-Verlag.

BROSS, U./KOSCHATZKY, K./STANOVNIK, P. (1999): *Development and Innovation Potential in the Slovene Manufacturing Industry*. Karlsruhe: Fraunhofer ISI (Arbeitspapier Regionalforschung Nr.16).

HERDEN, R. (1992): *Technologieorientierte Aussenbeziehungen im betrieblichen Innovationsmanagement*. Heidelberg: Physica-Verlag.

KALIN, T. (1999): *Future Institutional Changes in Research Environment in Slovenia: For Better or for Worse*. 4th Semmering Science & Technology Forum: Institutional Changes - Efficiency and Effectiveness; Interdisciplinary Centre for Comparative Research in the Social Sciences, Vienna, 3-5 December 1999.

KLINE, S.J./ROSENBERG, N. (1986): An overview of Innovation, LANDAU, R./ROSENBERG, N. (Eds.) *The Positive Sum Strategy*. Washington: National Academy Press, pp. 275-305.

KOMAC, M./KRAWCZYNSKI, J. (Eds.) (1994): *Conceptual Approaches to the Support of industrial Research and Development in Slovenia*. Jülich: Forschungszentrum.

KOSCHATZKY, K. (1999): Innovation Networks of Industry and Business-Related Services - Relation Between Innovation Intensity of Firms and Regional Inter-Firm Co-operation. *European Planning Studies*, 7, pp. 737-757.

KOSCHATZKY, K. (1998): Firm Innovation and Region: The Role of Space in Innovation Processes. *International Journal of Innovation Management*, 2, pp. 383-408.

KOSCHATZKY, K. (1997): Innovative Regional Development Concepts and Technology–Based Firms, KOSCHATZKY, K. (Ed.) *Technology-Based Firms in the Innovation Process. Management, Financing and Regional Networks*. Heidelberg: Physica-Verlag, pp. 177-201.

MESKE, W./MOSONSI-FRIED, J./ETZKOWITZ, H./NESVETAILOV, G. (Eds.) (1998): *Transforming Science and Technology System – the Endless Transition*. Amsterdam: IOS Press.

MEYER-KRAHMER, F./KUNTZE, U. (1992): Bestandsaufnahme der Forschungs- und Technologiepolitik, GRIMMER K./HÄUSLER, J./KUHLMANN, S./SIMONIS, G. (Eds.) *Politische Techniksteuerung*. Opladen: Leske/Budrich, pp. 95-117.

POTRATZ, W./WIDMAIER, B. (1999): *Frameworks for Industrial Policy in Central and Eastern Europe*. Aldershot: Ashgate.

SYDOW, J. (1992): *Strategische Netzwerke. Evolution und Organisation*. Wiesbaden: Gabler-Verlag.

WALTER, G.H. (1999): Innovation Agencies: Their Role in Science, Technology and Innovation Policy, PEJOVNIK, S./KOMAC, M. (Eds.) *FORUM BLED, Science and Technology Investment. Priority in the Development Strategy of the Nation*. Ljubljana: Ministry of Science and Technology, pp. 89-95.

WALTER, G.H./BROSS, U. (1997): The Adaptation of German Experiences to Building Up Innovation Networks in Central and Eastern Europe, KOSCHATZKY, K. (Ed.) *Technology-Based Firms in the Innovation Process. Management, Financing and Regional Networks*. Heidelberg: Physica-Verlag, pp. 263-286.

WALTER, G.H./HEYDRICH-RIEDL, E./OLESEN, K. (1997): *Development and Implementation of an Innovation Agency in the Republic of Slovenia*. Ljubljana, Inception Report I to the European Commission.

WALTER, G.H. (1992): *Integration einheimischer Hochschulen in die industrielle Modernisierung der Dritten Welt*. Karlsruhe: Rufdruck - Druck- und Verlagsgesellschaft.

WALTER, G.H. (1995): Slovene-German Co-operation in the Field of Technology Policy. PEJOVNIK, S./KOMAC, M. (Eds.) *FORUM BLED, International Scientific and Technological Co-operation: Problems, Challenges, Opportunities*. Ljubljana: Ministry of Science and Technology, pp. 69-75.

8. Innovation Networking in a Transition Economy: Experiences from Slovenia

Knut Koschatzky, Ulrike Bross

8.1 Problems of System Transformation and of the Development of Innovation Networks in Slovenia

At the beginning of the 1990s, due to the abolition of the socialist economic structure, the Central and Eastern European Countries (CEECs) were suddenly forced to thoroughly transform their political and economic order. This drastic situation is unique in history. With the transformation process, high expectations arose regarding policy, economy, and science. Due to the dissolution of the Yugoslavian Federation, Slovenian firms had to find new markets, and Slovenian research institutes had to alter their research priorities and intensify their contacts with industry (Walter/Bross 1997). Already in former Yugoslavia, Slovenia was the most industrialised region, which achieved 29 % of the total of Yugoslavian exports with only 8 % of the total population in 1990 (European Commission 1993: 5). Even after its independence in 1991,[1] the small country has remained primarily export-oriented with approximately 20 % of higher-priced technology products (pharmaceutical products, goods from the electronic and electrical engineering industry), approximately 50 % mid-tech and about 20 % low-tech products. Due to the process of economic adaptation, a drop in production was noted in all branches. Economic collapse especially threatened traditional industrial sectors such as steel and heavy machine construction, the automobile industry, and the textile and furniture sector. Although the transition of the industrial sector is not yet complete, the number of business enterprises has obviously grown during the past years, especially in the areas of small enterprises and the service sector.

---

1 On 15 January, 1992, the Republic of Slovenia was recognised as an independent state by the EU, and has been associated with the EU since 10 June 1996. In Slovenia, approximately 2 million inhabitants live on a surface of 20,256 km² (for a map of Slovenia cf. Figure 8-1). In 1996, the Gross Domestic Product (GDP) per capita reached 9,279 USD (EBRD 1997), the annual GDP growth rates were 3.1 % in 1996 and 3.8 % in 1998 (estimated); in 1997, the unemployment rate was 14.4 % (IMAD 1998; Raiser/Sanfey 1998). The largest towns are Ljubljana (276,000 inhabitants) and Maribor (108,000 inhabitants).

Proportionally to its small size, Slovenia's research potential is impressive: the number of students trained at the universities of Ljubljana and Maribor is 38,000; in addition, besides the Academy of Science and Arts, there are approximately 50 independent research institutes including the two most important ones, the Jozef Stefan Institute and the National Chemistry Institute. Already during the Yugoslavian period, Slovenia's scientific exchange with international partners was important. During the past years, several new institutions in the innovation infrastructure have been created. These are, for example, the Technology and Development Fund of the Slovenian Republic which, among other objectives, finances young technology firms and founders; several Technology and Business Incubator Centres were established at Ljubljana, Maribor, and Koper. However, at least until the mid-1990s, the readiness for co-operation both in the research sector and between science and industry remained insufficient in spite of the innovation policy led by the Ministry of Economy and Technology, which was oriented towards co-operation promotion; consequently, the potential of the Slovenian research and innovation sectors could not be utilised (Stanovnik 1998).

Figure 8-1:   **General map of the Republic of Slovenia**

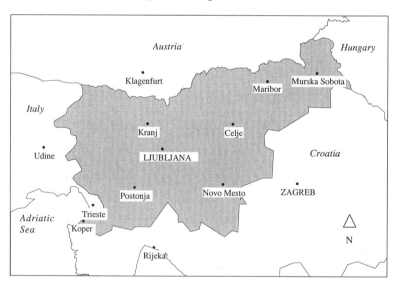

Due to these comparatively favourable framework conditions, Slovenia's technology and innovation policy is considered as exemplary in comparison with that of other Central and Eastern European Countries (e.g. Croatia, Slovakia, Hungary); consequently, it is seen as an example of an innovation system showing a definite approach of re-evaluating and re-structuring innovation activities and innovation-related network relationships, an approach resulting from the transition process which has taken place on a political, economic, and social level (Walter/Bross

1997). However, it must be clarified whether the pre-conditions for the development of innovation-oriented network relationships are present in transition countries, or if their socialist inheritance hinders their coming into existence.

A linear innovation model according to the soviet-leninist science push also predominated in the Slovenian part of Yugoslavia during the socialist period. Interactive learning processes and the feed-back of users' requirements were either underdeveloped or non-existent. This was also shown by the high degree of fragmentation which existed between the individual innovation institutions (Dyker/Perrin 1997; Meske 1998). These structural characteristics are linked to behaviours and routines which might have outlasted system transformation both in companies and in research institutes, and which now possibly hinder the co-operative organisation of innovation processes. The question is whether or not pre-conditions do exist for the development of innovation-oriented co-operation relationships in transition countries, or if the socialist history hinders their coming into existence.

Consequently, this contribution is based on the following assumptions:

- Due to a high degree of vertical integration, both innovation networks and the acquisition of complementary knowledge were less relevant during the socialist period than today.

- If external contacts existed, these were mostly related to production and sales (vertical relationships), with a strongly specialised knowledge exchange (fragmentation of innovation actors).

- For enterprises and research institutes it was almost or completely impossible to develop the ability to enter a horizontal knowledge exchange and realise appropriate learning processes. Horizontal co-operative relationships only had a complementary character in the socialist system, with the acquisition of material resources as their main role.

- Traditional structures and routines, e.g. those implemented in former public organisations and practised by executives without any (free market) experience abroad have survived the system transition.

- Most Slovenian research institutions are still strongly basic research oriented and are therefore not very well qualified to act as co-operation partners for innovation support and market oriented knowledge exchange with an industrial sector producing on an average technological level.

Based on these presumptions, Slovenia serves as a case study for an analysis of the entrepreneurial pre-conditions for innovation activities and innovation co-operation in a transition economy; moreover, the character and intensity of network relationships between firms and research institutes will be shown. This objective is not only of scientific interest but is also relevant for innovation policy. The promotional measures granted up to the present by the Directorates-General "Enterprise" (Directorate Innovation) and "Regional Policy" of the European Commission in view of the development and implementation of regional innovation strategies in European countries (cf. Landabaso/Youds 1999) will also be offered in a similar form to the accession countries of the European Union. As these measures are based on the concept of regional innovation systems and, consequently, on a close networking and co-operation system between regional actors (Braczyk *et al.* 1998; Cooke/Morgan 1998), this analysis of existing co-operation patterns between research institutions and companies can significantly contribute to the questions whether regional innovation systems do exist in transition economies and whether such measures should be applied in these countries. Therefore, the following research questions should be answered:

(1)   What is the significance of innovation-relevant co-operation relationships in Slovenia? With which partners do firms and research institutes co-operate, how frequently, and has the significance of external co-operation changed over time?

(2)   What kinds of firms are mainly integrated into networks? Is there a difference between vertical and horizontal innovation networks?

(3)   What kinds of co-operation relationships are predominant, and what is their spatial reach?

(4)   Which research institutes co-operate with companies, and in what way? What is their spatial co-operation pattern?

(5)   Which conclusions can be drawn for the existence and for the effects of innovation networks in Slovenia?

8.2   Basic Data

The data used in this analysis comes from a postal innovation survey carried out by the Fraunhofer Institute for Systems and Innovation Research (ISI), Karlsruhe, in co-operation with the Institute for Economic Research (IER), Ljubljana, among industrial companies, service companies, and research institutions, between October 1997 and March 1998.[2] In order to validate the postal survey, 27 additional interviews took place with industrial and service companies, research and transfer institutions, and Slovenian experts. The following analysis is based on the industry and the research institutions' sample.

---

[2]   This survey was part of the European Regional Innovation Survey (ERIS) carried out in 11 European regions by the Department of Economic Geography from Hanover University, the Chair of Economic Policy from the Technical University Bergakademie Freiberg, the Department of Economic and Social Geography of the University of Cologne, and the Fraunhofer Institute for Systems and Innovation Research, Karlsruhe, financially supported by the German Research Council (cf. Fritsch *et al.* 1998: 249; Sternberg 2000: 396-402).

Tables 8-1 and 8-2 show a comparison between the branch and size structure of both the total population and the sample of the industrial survey. The total population was identified from data received from the Slovenian Office for Statistics (SURS) about companies with 10 and more employees from the NACE categories 15-37.[3] A total of 1,336 questionnaires were distributed, 416 of these could be used for the empirical analysis, corresponding to a response rate of 31.1 %.

Table 8-1:   Composition of the industrial survey according to branches

NACE	Branch	Total population		Sample		
		Number of Firms	Percent	Number of Firms	Percent	Employees Percent
15, 16	Food, Beverages, Tobacco	118	8.8	33	7.9	7.6
17-19	Textiles, Clothing	160	12.0	46	11.1	11.4
20-22, 36	Wood, Paper, Printing	293	21.9	79	19.0	16.3
23-26	Chemical products, Plastics	194	14.5	54	13.0	12.2
27, 28	Metal processing	200	15.0	77	18.5	23.1
29, 34, 35, 37	Mechanical engineering, Vehicles	203	15.2	63	15.1	13.6
30-33	Electrical and optical equipm.	168	12.6	64	15.4	15.8
	Total	1,336	100.0	416	100.0	100.0

Source: ERIS Slovenia

Regarding the structure of branches and size of the companies, only slight, statistically insignificant differences were found between the total population and the sample; consequently, the sample may be considered to be a relatively representative image of the Slovenian companies which constitute the total population of the Slovenian manufacturing industry. The most important branch from a numerical point of view is the wood, paper and printing industry with a share of 19 % of the companies, followed by the metal processing industry with 18.5 %. Also in terms of employment figures, these two branches are the most important Slovenian industry sectors, of which metal processing clearly precedes the wood, paper and printing sector with a share of 23.1 %. The following positions are held by the electrical and optical equipment industry (15.4 % of the companies resp. a share of 15.8 % of all industrial employees), the mechanical engineering and vehicles industry (15.1 % resp. 13.6 %), and the chemical and plastics industry (13 % resp. 12.2 %). Consequently, Slovenia's industrial structure is dominated by two branches which are characterised by a relatively underdeveloped level of research and development activities in international comparison (Koschatzky/Traxel 1997: 27; Muller/Traxel 1997: 9). In Slovenia, this applies particularly to the wood, paper and printing industry, which spends 2 % of its turnover on average on R&D (median 1 %) (Bross

---

3   Survey units were enterprises, i.e. local production units including or in close proximity to administrative and auxiliary departments. For reasons of style, the terms "company", "enterprise", and "plant" are used as synonyms in this paper.

et al. 1999: 24). The industrial average in Slovenia is 3.8 % (median 2 %). In contrast to this, companies from the metal processing industry spend 5.2 % of their turnover on average on R&D; consequently, this branch holds the second position regarding R&D intensity, following the electrical and optical equipment industry with 6.5 %.

Table 8-2: **Composition of the industrial survey according to size classes**

Size classes (Employment)	Total population		Sample	
	Number of Firms	Percent	Number of Firms	Percent
1-19 Employees	319	23.9	79	19.0
20 – 49 Employees	326	24.4	83	19.9
50 – 99 Employees	225	16.8	78	18.7
100 – 199 Employees	220	16.5	83	20.1
200 – 499 Employees	169	12.6	55	13.2
500 – 999 Employees	49	3.7	22	5.3
> 1000 Employees	27	2.0	16	3.8
no information	1	0.1		
Total	1,336	100.0	416	100.0

Source: ERIS Slovenia

Regarding the size structure, the emphasis is on companies with up to 200 employees; 77.7 % of the companies of the sample firms fall within this area. On average, an industrial company in Slovenia has 235 employees (median 87). This level is clearly surpassed by the metal industry (average 293; median 46, however) but not reached by the wood, paper and printing industry (with an average of 202 employees at a median of 93). Large-scale enterprises with more than 500 employees hold a share of 9.1 %. Large-scale companies are found especially in the metal processing industry and in machine and automobile construction. In total, middle-class industrial structures are predominant in Slovenia, whereas the share of large-scale companies with more than 500 employees is obviously higher than, for example, in Saxony,[4] where the ERIS data shows only 2 % (Fritsch *et al.* 1996: 5).[5]

The research data record comprises 60 institutions (response rate 47.6 %); compared with the total population, the university institutes are slightly underrepresented, public research institutes and transfer offices are slightly over-represented. 50 % of the sample account for the latter, whereas private research institutes are

---

[4] In the following, the German Federal State of Saxony, one of the ERIS regions, will serve as a comparative example regarding certain aspects of co-operation, due to the fact that Saxony also experienced a new orientation of innovation networks after the political change.

[5] For further structural characteristics of the industrial sample, please refer to Bross *et al.* (1999).

represented with a share of 28.3 % and university institutes with a share of 21.7 % (cf. Table 8-3).[6]

Table 8-3: **Composition of the sample according to types of research institutes**

Type of research institute	Population		Sample	
	Number	Percent	Number	Percent
University institutes	38	30.2	13	21.7
Public research institutes and transfer offices	53	42.1	30	50.0
Other research institutes	35	27.8	17	28.3
Total	126	100.0	60	100.0

Source: ERIS Slovenia

No special emphasis can be found when examining the size structure of the institutes. Institutes with up to 10 researchers account for 30.5 %, institutes with more than 51 scientists account for 27.1 % (11-20: 22 %, 21-50: 20.3 %). Natural science and medicine are predominant research areas, where 37.9 % of all institutes focus on research and development activities, followed by economics and social science (22.4 %). While natural sciences are considered a typical research focus for Central and Eastern European Countries, in the areas of economics and social science the gap is currently being closed.

8.3 Co-Operation Patterns of Slovenian Companies

Innovation relevant co-operation is a very important information and knowledge resource for Slovenian companies. Of the 416 firms included in the industrial survey, 385 confirmed that their co-operation with at least one partner surpassed normal business relationships in the case of innovation relevant activities. This is a remarkably high number when considering that 318 companies (76.4 % of the sample) accomplished product or process innovation between 1994 and 1996. It can therefore be concluded that not all of the cited co-operative relationships are innovation-bound contacts, but that "normal" business relationships were also mentioned especially concerning co-operation with service companies.

Therefore, business-related service companies represent the most important co-operation partner; their role was emphasised by 80 % of the interviewed companies. Contacts with customers (72 %) and suppliers (56 %) reflect the importance of networks within the vertical value added chain. Co-operations which are characterised

---

6   We thank Daniela Rink for her assistance in analysing the research data.

by a high degree of freedom in partner search, are distinctly less significant for information and knowledge exchange: 37 % of the companies co-operate with research institutes, 27 % with other companies. As shown by these shares, there is hardly any difference between the co-operation behaviour of Slovenian companies and the co-operation patterns identified in other regions by the ERIS survey. In the case of Saxony, for example, co-operation partners are called on in a comparable sequence and with similar frequency (Fritsch et al. 1996: 21). Consequently, it is clear that vertical network relationships play a predominant role, but Slovenia does not represent an exception on this point.

The great significance of vertical co-operation is also shown by the changes which have taken place in external collaboration since 1991. 64 % of the companies have increased their contacts with customers and 58 % of them have increased their number of suppliers, whereas only 6 % and 5 % resp. of the firms have reduced their co-operation relationships. This increase shows the changes which have taken place in industrial co-operation since the decline of the socialist economic system. Contacts with business-related service companies have also been intensified, however less strongly, since only 38 % of the interviewed firms indicated such an increase in co-operation. An interesting fact is that former relationships with universities and industrial research institutions have been broken off since 1991. Whereas 11 % of the companies entered new co-operation relationships, 14 % of them gave up their contacts. Changes in the originally linear innovation model according to the soviet-leninist science push could have caused the decrease in co-operative behaviour. Industrial research requirements were apparently not always satisfied by the formerly mandatory co-operation relationships, so that discontent has led to a breaking off of former contacts. In addition, the economic transition process also brought with it significant reductions in the companies' R&D budgets, so that research institutes were called upon less often.

As has been shown by a variety of empirical studies, small enterprises are at a disadvantage in comparison with large-scale companies regarding the access to external knowledge (Koschatzky/Zenker 1999a). They have less personnel and financial resources to organise networks and to maintain effective network management. If they carry out R&D, then their innovation activity often focuses on the development of new technology solutions without being able to introduce them onto the market; where such complex innovation activities are necessary and where work has to be shared with other partners, such innovation activities are mostly carried out by larger companies working in close co-operation with external partners (Frisch 1993: 285). Due to familiarity with their spatial environment, as well as a supposed lesser risk regarding the control of regional network partners, the networks built up by smaller companies are oriented towards their closer spatial environment. Moreover, small and medium-sized companies without own R&D capacities have limited knowledge potential at their disposal, which, however, could ensure access to research networks (Rosenberg 1990: 170; Hicks 1995). For these reasons such companies only practice co-operation in a limited way.

This fundamental pattern of industrial co-operation behaviour is confirmed in the case of Slovenia (cf. Figure 8-2). The share of small companies with up to 100 employees maintaining external innovation co-operation, is far below that of companies with more than 100 employees. This is true for all three of the groups of co-operation partners; with a strikingly clear increase in research co-operation for companies with more than 100 employees. Moreover, an effect of size is seen regarding the frequency of co-operation in view of their export orientation, i.e. their presence on international markets. Contacts with customers, suppliers and research partners increase on average up to an export share of 75 %; and beyond this, most of them remain on the same level. The necessity for external co-operation in view of the development and adaptation of internationally competitive products is shown by the increasing shares, especially for those companies which depend on international demand.

Figure 8-2: **Co-operation with different partners according to employment size-classes and export shares**

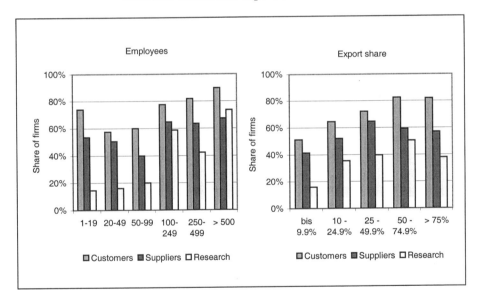

Source: ERIS Slovenia

Although statistically insignificant, a difference in co-operation behaviour is also found between the branches of the manufacturing industry. In Table 8-4, those companies which carried out innovation projects between 1994 and 1996 are distinguished from the total. It can also be seen in the case of Slovenia that innovating companies show more tendency for co-operation than all companies on average. Whereas, in comparison with the average co-operation frequency, fewer companies of the food processing industry have an information and knowledge exchange with their customers (probably due to the high number of final consumers), contacts with suppliers and information exchange with research institutions play an above-average role in this sector. The most intense co-operation with all three of the groups of partners is found in the chemical and plastics industry and the electrical and optical equipment industry. Both branches also excel with the highest intensity of scientific contacts. In the light of the generally decreasing number of co-operation activities with research institutions, this could point to a demand-oriented research offer for these companies, as well as to a high absorptive capacity due to the high technological standard in these sectors.

Table 8-4:   **Co-operation partners according to branches**
(share of firms in %)

Co-operation partners Branches	Customers		Suppliers		Research institutes	
	all firms	innov. firms	all firms	innov. firms	all firms	innov. firms
Food, Beverages, Tobacco	57.6	66.7	63.4	74.1	42.4	48.1
Textiles, Clothing	71.7	91.2	56.5	67.6	30.4	38.2
Wood, Paper, Printing	63.3	80.4	49.4	62.5	25.3	33.9
Chemical products, Plastics	75.9	88.1	61.1	73.8	50.0	57.1
Metal processing	76.6	92.7	49.4	60.0	31.2	41.8
Mechanical eng., Vehicles	74.6	87.5	55.6	68.8	27.0	33.3
Electrical and optical equipm.	78.1	85.7	62.6	66.1	53.1	57.1
Total	71.9	85.5	55.8	66.7	36.1	44.0

Source: ERIS Slovenia

The increasing tendency towards co-operation on the side of companies in accordance with the increase of R&D or innovation intensity, is also shown by the following two input indicators. The number of employees holding a university degree (or a comparable qualification) is an indicator for both the knowledge intensity and the absorptive capacity of companies. R&D intensity, which is measured by the turnover percentage spent on R&D, reflects a company's R&D commitment in monetary terms. Often R&D intensity is also used to classify companies in Low-Tech (up to 3.5 % R&D intensity), Medium-Tech (3.5-8.5 %), and High-Tech companies (more than 8.5 %) (Gehrke/Grupp 1994: 40). Both with the rising share of highly qualified employees and the increasing R&D intensity, the share of companies with co-operative relationships is increasing, whereby there is an obvious link between the frequency of research contacts and an increased number of qualified employees (cf. Figure 8-3).

Figure 8-3: **Innovative co-operation with external partners according to input factors**

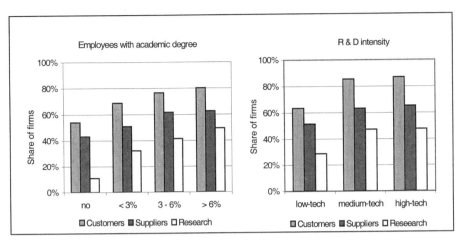

Source: ERIS Slovenia

In contrast to the characteristics used for the differentiation of industrial co-operation intensity so far, transition-specific aspects are shown by the types of ownership of the companies, as well as by their date of privatisation. The most intensive network connections with partners (in particular with customers) are maintained by Slovenian public enterprises, which is certainly also due to the size of these companies. Only gradual differences exist between the remaining types of ownership (social ownership, foreign ownership, private companies with Slovene owners) especially regarding co-operation activities with research institutions, which, for their part, have co-operation with a few foreign companies and private Slovenian companies. Consequently, the type of ownership does not significantly influence the choice of partners for innovation networking. The most important co-operation shares are shown by companies which were privatised by 1992, whereas a lower degree of co-operation intensity is shown by firms which have only recently been privatised, or which have not yet been privatised. This pattern is due to several reasons. In this way, competitive companies with intense external co-operation relationships, were the first to be privatised. Moreover, in comparison with companies which were only recently privatised, they have had more time to establish new contacts and gather experience with innovation networks.

8.4   Types of Co-Operation

Different functions can be fulfilled by innovation networks. They enable access to information, for which informal contact is usually sufficient. The more the relationships are directed towards joint R&D projects or the market introduction of new products, the more they have to be formal (and reliable). R&D-oriented networks especially allow for learning processes, since not only information but also knowledge and competence are exchanged.

The general exchange of information, the generation of new ideas, and the development of concepts are the characteristics of relationships mostly organised on an informal level, whereas the development of prototypes, pilot applications, and market introduction represent formal co-operation relationships. Informal contacts between all partners are predominant in Slovenian networks, and here it is also obvious that co-operation relationships in Slovenian industry exist mainly with customers and service companies. Regarding customers, increasingly formalised co-operation relationships have, for the present, led to a drop in the number of intense or very intense co-operation partnerships although, finally, the ranking of market introduction reached a level of importance as high as for general information exchange. This reflects the sales-oriented contacts with this industrial group, in which innovations are realised in close co-operation and with respect of the market. Suppliers' networks are primarily oriented towards general information exchange and the generation of new ideas. Compared with customer relationships, co-operation relationships aimed at market introduction play a distinctly minor role. Service companies play a predominant role as information partners, whereas the remaining aspects of innovation are obviously less important. Not only do research institutions have a minor position as co-operation partners; they are also only considered to be supportive in the early phases of the innovation process. They are only called on in a limited way for the joint development of prototypes, for pilot applications or even for market introduction. This low level of utilisation might point to an insufficient qualification of the institutes regarding innovation support close to the market. However, an analysis of the research co-operation from the point of view of the institutes should be consulted before drawing more serious conclusions (cf. section 8.6).

The type of co-operation relationships shows that formal aspects of co-operation only play a significant role in vertical networks with customers (market introduction). Otherwise, informal contacts are predominant. Also on this point, Slovenian industrial companies only differ slightly from those of other regions. In Saxony, for example, a similar structure of co-operation types is found (Fritsch *et al.* 1996: 24-25). The comparison of further regions will show whether this is a universally valid character of networks, or whether transition has led to cautious behaviour with external partners.

8.5   Spatial Reach of Co-Operation Relationships

Sufficient access to information and knowledge is only assured if companies maintain complementary networks, i.e. if they can access both regional and interregional networks (Camagni 1994). Of the three co-operation partners customers, suppliers, and research institutions, Figure 8-4 shows the shares of companies which maintain contacts with their respective partners from Slovenia and the neighbouring countries (Hungary, Austria, Italy, Croatia), as well as from other foreign countries. A similar spatial structure of both the co-operation relationships with customers and those with suppliers is found. Approximately 38 % of the companies co-operate with customers and suppliers from Slovenia, 29 % with the corresponding partners from the neighbouring countries, and 33 % with companies from other foreign countries. The spatial openness of vertical networks is due to the limited market potential available in a country of the size of Slovenia, which necessitates co-operation with foreign partners. Even if the neighbouring countries were considered with the region of Slovenia for reasons of comparability, the remaining share of foreign contacts is distinctly higher than that of Saxonian companies with only approximately 7 % of their co-operating customers and suppliers being foreign (Fritsch *et al.* 1996: 27). In the same way, in Baden, another ERIS region, co-operation partners which are located abroad only play a comparably minor role, whereas Alsatian companies, with an average share of 25 % of foreign customers and suppliers, are oriented towards long-distance contacts in about the same way as Slovenian firms (Koschatzky 1998: 284).

Figure 8-4:   **Spatial reach of co-operation relationships of Slovenian companies**

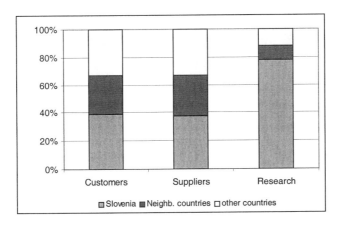

Source: ERIS Slovenia

A completely different pattern is seen in the area of research co-operation: here, national institutions predominate with a share of almost 80 %. Only 10 % of the companies co-operate with institutes from neighbouring countries and the remaining foreign countries. As has already been shown, it is true that networks with research institutions rank distinctly lower than vertical relationships; however, if appropriate contacts are established, then they are built with national institutions. Consequently, an insufficient Slovenian research offer cannot be the only reason for the low level of co-operation intensity; in this case, companies would take more advantage of foreign institutes. On the one hand, this co-operative behaviour could be due to routine (former negative experience is projected onto the present); on the other hand, it might be due to companies' lack of capacity and competence to establish contacts with research institutions. Companies which do co-operate with research institutes seem to use them as a bridgehead which guarantees access to internationally available information and knowledge. Such behaviour is also found in other regions of the ERIS database (e.g. Saxony, Baden), and does not represent a Slovenian particularity.

Where a relationship is built up between different industrial characteristics and the spatial reach of innovation networks, the following is found:

- There are no explicit differences between *branches*; these are only significant regarding the relationships with customers and suppliers from the remaining foreign countries, including EU countries. In this way, electrical and optical equipment producing companies especially co-operate with customers and suppliers from the remaining foreign countries, whereas supplier networks of the mechanical engineering and vehicles industry are restricted to Slovenia and its neighbouring countries. Companies from the chemical and plastics industry co-operate closely with national customers, whereas, due to the comparably high Slovenian wage level, the textile and clothing industry concentrates on suppliers of intermediate products from neighbouring and other countries.

- Related to the *size of companies* (employment figures), an increase in international co-operation relationships (with neighbouring and other countries) is found parallel to an increase in size. This shows that the capacity to organise networks and to establish contacts with international partners grows with an increase in company size.[7] Whereas the share of companies co-operating with customers located in Slovenia slightly decreases with an increase in company size, no such effect of size is found in the suppliers' networks. From this angle, no change of the spatial co-operation pattern is found in horizontal networks due to altered company sizes; the reason for this is the great significance Slovenian research institutions have as network partners.

---

7 Here, company size is not only seen as a quantitative variable, but as an indicator of various structural features which characterise companies of different sizes. One of these features is their absorptive capacity, which is increased due to manifold resources with growing firm size (Koschatzky/Zenker 1999a).

- Regarding *R&D intensity* of the companies, a minimal difference is found as in the spatial reach of vertical and horizontal innovation networks of the different sectors. No relationship is found between increased R&D intensity and a widened spatial reach of networks. However, companies from the R&D intensity class reaching from 3.5 % - 8.5 % co-operate more frequently with customers from other countries than with their respective national partners. Otherwise, co-operation partners from Slovenia hold the first place, followed by companies and research institutions from neighbouring and other foreign countries.

- A differentiation according to the companies' *type of ownership* only reveals significant differences for few of the three groups of spatial reach. Nevertheless, a predominant tendency shows that both foreign companies and state-owned Slovenian companies maintain an above-average frequency of international contacts with customers, suppliers and research institutions, whereas private Slovenian companies are more intensely integrated into their closer spatial environment. An even more intense regional orientation regarding innovation networks is shown by socially owned companies, the prevailing business type until independence.[8] Their customer and research contacts especially are oriented towards Slovenia, whereas relations with suppliers correspond more or less to the average pattern of all other companies. This co-operation behaviour points to more openness in the use of internationally available knowledge on the side of foreign companies and Slovenian public firms, whereas private firms and the still socially owned companies prefer spatial closeness for their information and knowledge exchange. This is also due to size effects showing that increased performance and absorptive capacities of larger companies (e.g. state-owned companies) lead to a wider spatial reach of innovation-relevant co-operation relationships.

8.6 Innovation Co-Operation with Companies from the Point of View of Research Institutions

When considering all Slovenian research and transfer institutions, companies from the production and service sector represent the most important co-operation partners. 82.5 % of the sample institutions co-operate with firms, 80.7 % co-operate with other research institutions. The successive positions are held by public administration (59.6 %), trade associations (49.1 %), transfer institutions (43.9 %), research institutions which participate in common research programmes promoted by the EC (42.1 %), as well as banks and financial institutes (15.8 %). Besides contacts with other research institutions, this distribution shows a strong orientation towards the business sector by the Slovenian research institutes, which was not revealed as such by the industrial survey. This could be due to the fact that the survey showed

---

8 Privatisation of these companies has not yet been concluded.

the co-operation behaviour of companies from the manufacturing industry, whereas the comments of the research institutes referred to production and service companies. Furthermore, the government (public administration) still plays an important role regarding both the country's scientific exchange and the financing of universities and public research institutions.

On the whole, co-operation with companies has improved since independence in 1991: this was confirmed by 36.2 % of all research institutions. For more than half of them, co-operation has remained the same, and only 12.1 % consider it as worse than before. Also from the point of view of the other co-operation partners, an improvement in co-operation relationships predominates in contrast to a deterioration. In this way, research institutes consider the development of their external contacts as clearly more positive than industrial companies do.

Distinguishing between university institutes, public R&D institutions or transfer agencies and other, mainly private research institutes, differences become obvious concerning the intensity of co-operation shared with companies and other research institutions (cf. Table 8-5). Although these are not statistically significant due to the small number of cases, they point to the fact that university institutes co-operate less intensely with companies than the other two groups of research institutions. This is partly due to the predominance of the educational role played by universities in the socialist system, so that research capacities had to be improved following the country's independence. The closest co-operation exists between Slovenian university institutes and the public administration (66.7 % indicated intense or very intense co-operation), which reveals that university research activities are still closely related to the public sector. In contrast to this, other, mainly private institutes are clearly more oriented towards industry. They co-operate principally with enterprises, while only 23.5 % co-operate intensely with the public administration. However, this closeness is relativised by the fact that only a few of these institutes maintain intense contact with trade associations and financial institutions. The data does not indicate whether reservedness regarding co-operation predominates on the side of the trade associations and banks, or on the side of the institutes. As far as the mixed group of public R&D and transfer institutes is concerned, they also show an above-average orientation towards both companies and other research institutions; therefore, they are more intensely embedded in innovation networks with these partners than the other organisations.

Table 8-5: **Intensity of research institutes' co-operation with firms and other partners**

(shares in %)

Type of institute	Co-operation with...					
	Firms		Institutes		Public Admini-stration	
	low	intensive	low	intensive	low	intensive
University institute (n=12)	58.3	41.7	50.0	50.0	33.3	66.7
Public research institute/ transfer office (n=30)	36.7	63.3	40.0	60.0	50.0	50.0
other research institute (n=17)	41.2	58.8	70.6	29.4	76.5	23.5

Source: ERIS Slovenia

Regarding research institutions, the tendency towards co-operation has also in-creased parallel to growth, which is comparable with industry. This is true both for co-operation relationships with industry and with other research institutions. Con-cerning co-operation relationships with industry, this size effect is statistically sig-nificant (5 % level). Regarding research co-operation, it is less pronounced due to the fact that small institutes frequently co-operate with other institutes. Referring to the institutes' disciplines, mechanical engineering has the largest share of co-operation with industry; all of the institutions of this discipline co-operate with in-dustry. With a co-operation share of 81.8 %, the second place is held by the natural sciences and medicine. Economics and the social sciences, as well as electrical en-gineering, do not reach the Slovenian average for institutes co-operating with in-dustry of 70.7 %.

As shown in Figure 8-5, research institutions assess co-operation with industry in a slightly different way from companies. The most important element of co-operation is the development of prototypes, during which 38 of the 53 institutions (71.7 %) worked intensely to very intensely with companies from the industry and service sector. This is reflected by the production-oriented competence on the side of the institutes, attained during the socialist period; many of the institutes had even cre-ated their own small production units in order to earn money. Other important as-pects of co-operation are conceptional work and the generation of new ideas. In comparison with these, general information exchange, pilot applications, or co-operation in view of market introduction play a slightly less important role. Conse-quently, as seen from the perspective of all industrial co-operation partners of re-search institutions, formal relationships hold a high position. Due to the fact that formal relationships often grow through informal contacts and the mutual trust

gained therein (Beise/Stahl 1999: 410), research institutions have been able to develop such connections based on trust with their partners from industry to a higher
degree than companies did up to then. This is also shown by "market introduction"
as an objective of co-operation, which was only relevant for few industrial companies. In contrast to this, no less than 37.7 % of the institutes indicated intense or
very intense co-operation aimed at market introduction. Especially for this kind of
co-operation, the transfer of implicit or at least sensitive knowledge must be presumed. Together with the high share of national scientific co-operation, seen from
an industrial perspective, this result is at least a weak prove for the significance of
spatial proximity for the exchange of implicit and trust-based knowledge.

Figure 8-5: **Kinds of co-operative relationships of research institutes
with firms**

(absolute number of institutes)

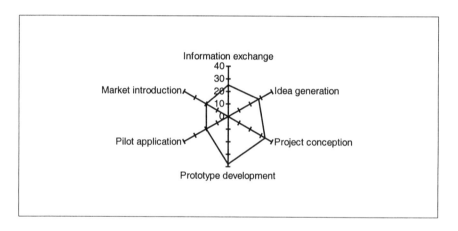

Source: ERIS Slovenia

A slightly different picture is found when looking at the three sub-groups of institutes. It is shown that, with the exception of the co-operation target "market introduction", other R&D institutions have less intense co-operation with companies
than the remaining institutes (cf. Figure 8-6). Surprisingly, this aspect of innovation
seems to be one of the strong points of university institutes; in fact, the share of
university institutes which have market-close co-operation with companies aimed at
innovation is higher than the share found for all other institutions and for all other
regions.[9] Due to the limited number of Slovenian cases, this result should not be
over-estimated; however, it points to the fact that some university institutes have

---

[9] In Saxony, for example, the 22 % share of private research institutions which co-operate with
   companies in order to realise ideas which have already been developed, is higher than the 17 %
   percentage shown by universities (Fritsch *et al.* 1997: 23).

market related qualifications in spite of their generally limited number of industrial contacts. In contrast to this, the involvement of public R&D institutions in market introduction is only 29.6 %; they are more specialised in basic research and, consequently, offer only limited support for this kind of co-operative relationship.

Figure 8-6: **Kind of co-operative relationships with firms according to type of institute**

(shares of institutes in %)

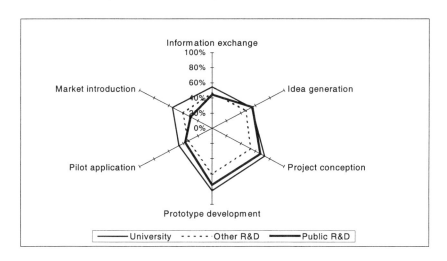

Source: ERIS Slovenia

One of the roles of research institutions is to serve as a bridgehead for small and medium-sized companies, the co-operative behaviour of which is mostly limited to their regional environment (Feldman 1994; Koschatzky/Zenker 1999a), by providing internationally available information and scientific know-how; in order to accomplish this function, they must have access to international knowledge networks. This points to the spatial reach of co-operation with research institutions in one respect, and with firms in another. Figure 8-7 shows the numbers of research institutions intensely co-operating with other institutions from Slovenia, from the neighbouring countries Italy, Croatia, Hungary and Austria, as well as from other European and non-European countries. It is shown that especially the other, non-governmental R&D institutions of the sample are not able to fulfil a science-oriented bridgehead function. Their scientific co-operation is almost exclusively oriented towards Slovenia. Noteworthy international links are non existent. University and public R&D institutions are subject to different conditions.[10] About 40 %

10 Due to their status, the so-called "national institutes" benefit from higher basic funding and have good chances of being granted state assistance for projects. In Central and Eastern European

of them are integrated into European and world-wide knowledge networks; public institutes are slightly more oriented towards Slovenian regions than university institutes are. Consequently, these two groups should be able to make international knowledge accessible to companies. On the one hand, university institutions have the potential for this kind of positive network effects; on the other hand, due to the fact that they co-operate less with firms than the other two groups (cf. Table 8-5), this potential is not yet sufficiently known to and made use of by companies.

Figure 8-7: **Spatial pattern of research co-operation according to type of institute**

(share of co-operating institutes in %)

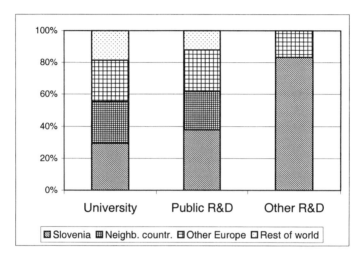

Source: ERIS Slovenia

Regarding co-operation with companies, both university and public research institutes focus on Slovenian co-operation partners (cf. Figure 8-8). This share is approximately 55 % for the university institutes and more than 70 % for public organisations. This confirms the results of the industrial survey for both of the groups, according to which industrial relationships are organised over shorter spatial distances. Consequently, not only public institutes but also university institutes fulfil the function of a national bridgehead by providing Slovenian companies with their own knowledge and with knowledge acquired on an international level. On the other hand, it is also possible to conclude that Slovenian university institutes and public research institutes do not represent interesting co-operation partners for many foreign companies, since their services are only called upon in a limited way. Pri-

Countries these financing sources represent an important pre-condition for the establishment of international co-operation relationships.

vate institutes show different co-operation behaviour; their innovation co-operation with industrial partners has a distinctly more international character. Due to the fact that they are less integrated in scientific networks, companies are their most important co-operation partners and their most important sources of knowledge. Innovation-relevant knowledge, which qualifies them as interesting co-operation partners, can be accumulated not only through scientific networking but also through learning processes which are induced by industrial co-operation.

Figure 8-8:   **Spatial pattern of firm co-operation according to type of institute**

(share of co-operating institutes in %)

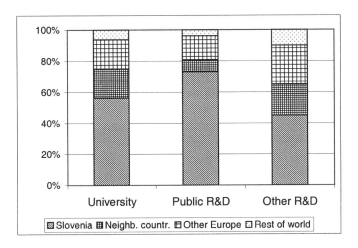

Source: ERIS Slovenia

8.7   Summary and Conclusions for Innovation Policy

To the research questions formulated in chapter 8.1, the following answers result from the empirical analysis:

(1)   Slovenian companies show the same level of external co-operation relationships as companies from Central European regions. This represents a predominance of vertical co-operation relationships with customers and subcontractors as was found in the compared regions. Vertical networks have gained importance since the system change, in comparison to a distinctly smaller gain in the importance of horizontal networks with service companies and research institutions. Co-operation with research institutions represents a special case; apparently, during the socialist period, these could not prove to

be reliable partners, meeting all industrial requirements; consequently, after 1991, relationships with them were broken off by many companies.

(2) Slovenian firms show a tendency to increased co-operation activity parallel with company size, with a distinct increase in scientific co-operation for companies with more than 100 employees. Size effects also exist regarding the export share, the share of qualified employees, and the R&D intensity. Herewith, the primary significance of vertical co-operation (customers, suppliers) and the secondary significance of horizontal co-operation (research institutions) remains unchanged. Higher shares of co-operation relationships (especially with customers) are shown by companies which were privatised a longer time ago, as well as by state-owned companies. Whether former sales relationships were maintained in the first case was not answered by the data. It is presumed that the companies which were privatised at an earlier time distinguish themselves by better performance and more intense network relationships, whereas external relationships first have to be established or stabilised by companies which have only recently been privatised.

(3) Slovenian companies primarily use innovation networks for information exchange organised on an informal level. No note-worthy learning processes can be expected from the realisation of R&D or innovation projects in close co-operation. Only in vertical networks companies accept formalised co-operation with customers aimed at market introduction. Due to the size of Slovenia, companies depend on foreign co-operation partners. More than 60 % of the firms indicated co-operation with customers and suppliers from the neighbouring countries and the remaining foreign countries. The existence of relationships with Hungarian and Croatian companies dating from the socialist area is presumed at least as far as contacts with the neighbouring countries are concerned. Despite this, the high share of co-operation partners from the remaining foreign countries points to a strong orientation of Slovenian industry towards foreign countries, which is explained by the necessity of a presence on international markets, and which also has a long tradition. Whereas, regarding the spatial reach of innovation networks, only slight differences exist between branches or even companies with different R&D intensity,[11] the share of network relationships with foreign partners shows a clear increase due to growing company size. This is not a Slovenian particularity; rather, it shows that it is easier for large-scale companies than for small firms to establish and maintain international networks. At least partially, the size effect also plays a role regarding the companies' type of ownership. Larger public firms and foreign companies are more oriented towards long-distance network relationships than private Slovenian companies. These, however, distin-

---

[11] At least in the case of Slovenia, this allows for the conclusion that no causal relation exists between the increase in R&D intensity and the international character of innovation co-operation.

guish themselves from the companies which have not yet been privatised and which are even more bound to Slovenia for innovation relevant information.

(4) Company-related co-operation relationships are especially maintained by public and private R&D institutions. University institutes show intense co-operation with public administration and are represented in scientific networks in the same way as public R&D institutions. However, private R&D institutes only have minimal co-operation with other research institutes. Although university institutes co-operate less with companies than other institutions, the main emphasis of this co-operation is the support of market introduction. Consequently, knowledge potential, although present at university institutes, is apparently not sufficiently emphasised in contacts with companies. In total, the share of formal contacts with companies is distinctly higher than that of company-close innovation networks with research institutions; this can be seen as an indication of actually existing knowledge exchange, and evaluated as realised learning processes between science and economy. The close contact between Slovenian research institutions and companies is also shown by the spatial reach of innovation networks. While universities and public institutes in particular maintain co-operation relationships in their regional environments, they are also integrated into international scientific networks. Consequently, they have the potential to make international scientific know-how available and utilisable for industry. Private research institutions face a different situation. Although working mainly with Slovenian institutes, their frequency of co-operation with foreign companies is distinctly higher. Their relationships seem to focus on practical aspects of the innovation process, for which industrial experience is more important than scientific knowledge.

In order to carry out innovations, companies rely both on the presence of internal innovation capacity and on co-operation with external partners. In Slovenia, this is particularly true for vertical relationships with customers as well as, though to a lesser degree, with suppliers. As the most important vehicles for co-operation, Slovenian companies indicated access to resources, the minimisation of uncertainties, and the entrance to new fields of technology. The empirical analysis revealed that the chances of common interactive learning processes and of implicit knowledge transfer are not fully utilised, despite a relatively high share of co-operating companies at first glance. Co-operation takes place firstly on the level of information exchange and less so on the level of joint research and development.

Compared with vertical co-operation, Slovenian industry and research show a lack of co-operation, which is a typical pattern for all Central and Eastern European transition countries. Co-operation with foreign research institutes, in particular, is almost non-existent. However, integration into global research networks and international industrial networks is indispensable for gaining new knowledge. In this area, at least a partial docking function is assumed by Slovenian research institutes.

Innovation networks seem to be an appropriate model for the support of the restructuring process of regions and industries, based on existing technology competence and potential. For the initiation of concrete networks, an important role is played not only by private and public participants but above all by regional actors. However, the experience gained with the establishment and expansion of regional innovation networks in a fully developed free market economy cannot simply be applied to a post-socialist transition economy. Instead, experience must be adapted to the structure of the innovation process pursued in the transition country. In this way, innovation can be based on the application of scientific findings in industry or on a new combination of technological know-how.

Different political conclusions come out of the preceding analysis. Regarding the Slovenian example, it must be queried for which industries a "technology transfer" from the scientific sector seems important, or, which industries need support in particular through industrial co-operation. Both the technology paradigm as a general basis on which industry is based, and the absorptive capacity of firms are important criteria for the introduction of political measures. However, changing innovation networks in developed free market economies also point to the fact that continuing competitiveness is not guaranteed. On the contrary, also for transition countries which are able to master a "catching up process" in comparison with developed free market economies, it is true that new technology must continuously be further developed and spread, if they do not want to fall back in the international innovation competition. Essential pre-conditions therefore are the openness of innovation networks and the adaptation of national and regional strategies of innovation policy. Neither developed free market economies nor former transition countries are spared this continuous process of adaptation; however, in contrast to the former socialist system, the newly developing economic systems should enable this adaptation process more easily.

8.8   References

BEISE, M./STAHL, H. (1999): Public research and industrial innovations in Germany, *Research Policy*, 28, pp. 397-422.

BRACZYK, H.-J./COOKE, P./HEIDENREICH, M. (Eds.) (1998): *Regional Innovation Systems. The role of governances in a globalized world*. London: UCL Press.

BROSS, U./KOSCHATZKY, K./STANOVNIK, P. (1999): *Development and Innovation Potential in the Slovene Manufacturing Industry. First analysis of an innovation survey*. Karlsruhe: Fraunhofer ISI (Working Papers Firms and Region R1/1999).

CAMAGNI, R. (1994): Space-time in the concept of "milieu innovateur", BLIEN, U./HERRMANN, H./KOLLER, M. (Eds.) *Regionalentwicklung und regionale Arbeitsmarktpolitik. Konzepte zur Lösung regionaler Arbeitsmarktprobleme?* Nürnberg: IAB, pp. 74-89 (Beiträge zur Arbeitsmarkt- und Berufsforschung 184).

COOKE, P./MORGAN, K. (1998): *The Associational Economy. Firms, Regions, and Innovation.* Oxford: Oxford University Press.

DYKER, D./PERRIN, J. (1997): Technology policy and industrial objectives in the context of economic transition, DYKER, D. (Ed.) *The Technology of Transition. Science and Technology Policies for Transition Countries.* Budapest: Central European University Press, pp. 3-19.

EBRD [EUROPEAN BANK FOR RECONSTRUCTION AND DEVELOPMENT] (1997): *Transition Report 1997. Reformbericht Osteuropa, Baltikum, GUS.* Bonn: Lemmens Verlags- & Mediengesellschaft.

EUROPEAN COMMISSION (UNAVE/SEPSU) (1993): *Science and Technology in Slovenia. Final Report (1st Draft).* Brussels: European Commission.

FELDMAN, M.P. (1994): Knowledge Complementarity and Innovation, *Small Business Economics*, 6, pp. 363-372.

FRISCH, A.J. (1993): *Unternehmensgrösse und Innovation. Die schumpeterianische Diskussion und ihre Alternativen.* Frankfurt: Campus-Verlag.

FRITSCH, M./BRÖSKAMP, A./SCHWIRTEN, C. (1996): *Innovationen in der sächsischen Industrie – Erste empirische Ergebnisse.* Freiberg: TU Bergakademie Freiberg (Freiberger Arbeitspapiere 96/13).

FRITSCH, M./BRÖSKAMP, A./SCHWIRTEN, C. (1997): *Öffentliche Forschung im Sächsischen Innovationssystem – Erste empirische Ergebnisse.* Freiberg: TU Bergakademie Freiberg (Freiberger Arbeitspapiere 97/2).

FRITSCH, M./KOSCHATZKY, K./SCHÄTZL, L./STERNBERG, R. (1998): Regionale Innovationspotentiale und innovative Netzwerke, *Raumforschung und Raumordnung*, 56, pp. 243-252.

GEHRKE, B./GRUPP, H. (1994): *Innovationspotential und Hochtechnologie. Technologische Position Deutschlands im internationalen Wettbewerb.* Heidelberg: Physica-Verlag (2nd edition).

HICKS, D. (1995): Published Papers, Tacit Competencies and Corporate Management of the Public/Private Character of Knowledge, *Industrial and Corporate Change*, 4, pp. 401-424.

IMAD [INSTITUTE OF MACROECONOMIC ANALYSIS AND DEVELOPMENT] (1998): *Slovenian Economic Mirror*, No. 4. Ljubljana.

KOSCHATZKY, K. (1998): Innovationspotentiale in grenzüberschreitender Perspektive. Die Regionen Baden und Elsass, *Raumforschung und Raumordnung*, 56, pp. 277-287.

KOSCHATZKY, K./TRAXEL, H. (1997): *Entwicklungs- und Innovationspotentiale der Industrie in Baden. Erste Ergebnisse einer Unternehmensbefragung.* Karlsruhe: Fraunhofer ISI (Arbeitspapier Regionalforschung No. 5).

KOSCHATZKY, K./ZENKER, A. (1999a): *The Regional Embeddedness of Small Manufacturing and Service Firms: Regional Networking as Knowledge Source for Innovation?* Karlsruhe: Fraunhofer ISI (Working Papers Firms and Region R2/1999).

KOSCHATZKY, K./ZENKER, A. (1999b): *Innovative Regionen in Ostdeutschland – Merkmale, Defizite, Potentiale.* Karlsruhe: Fraunhofer ISI (Arbeitspapier Regionalforschung No. 17).

LANDABASO, M./YOUDS, R. (1999): Regional Innovation Strategies (RIS): the development of a regional innovation capacity, *SIR-Mitteilungen und Berichte,* Bd. 17, pp. 1-14.

MESKE, W. (1998): Toward New S&T Networks: The Transformation of Actors and Activities, MESKE, W./MOSONI-FRIED, J./ETZKOWITZ, H./NESVETAILOV, G. (Eds.) *Transforming Science and Technology Systems – the Endless Transition.* Amsterdam: IOS Press, pp. 3-26.

MULLER, E./TRAXEL, H. (1997): *Entwicklungs- und Innovationspotentiale der Industrie im Elsass.* Karlsruhe: Fraunhofer ISI (Arbeitspapier Regionalforschung No. 8).

RAISER, M./SANFEY, P. (1998): Statistical review, *Economics in Transition,* 6, pp. 241-286.

ROSENBERG, N. (1990): Why do firms do basic research (with their own money)?, *Research Policy,* 19, pp. 165-174.

STANOVNIK, P. (1998): The Slovenian S&T transition, MESKE, W./MOSONI-FRIED, J./ETZKOWITZ, H./NESVETAILOV, G. (Eds.) *Transforming Science and Technology Systems – the Endless Transition.* Amsterdam: IOS Press, pp. 98-107.

STERNBERG, R. (2000): Innovation Networks and Regional Development - Evidence from the European Regional Innovation Survey (ERIS): Theoretical Concepts, Methodological Approach, Empirical Basis and Introduction to the Theme Issue, *European Planning Studies,* 8, pp. 389-407.

WALTER, G.H./BROSS, U. (1997): Transformation deutscher Erfahrungen beim Aufbau von Innovationsnetzwerken in Mittel- und Osteuropa, KOSCHATZKY, K. (Hrsg.) *Technologieunternehmen im Innovationsprozess. Management, Finanzierung und regionale Netzwerke.* Heidelberg: Physica-Verlag, pp. 267-291.

9. Integration through Industrial Networks in the Wider Europe: An Assessment Based on Survey of Research[*]

Slavo Radosevic

032

038

F15 P23

F21 P33

9.1 Background

Integration of the central and eastern European countries (CEECs) through the Single Market and conformance to *acquis communitaire* dominate the policy agenda of the EU enlargement. Integration through different forms of inter-firm co-operation has been considered to be unproblematic and the automatic outcome of institutional and policy integration. It is assumed that the deeper institutional integration the more likely that it will lead to more extensive industrial co-operation.

The underlying rationale for this paper is the assumption that integration through industrial networks will not take place automatically as an effect of institutional integration. Although market integration is a necessary objective of enlargement, it is in no way a sufficient condition of dynamically efficient outcomes for an enlarged EU. Convergence of the CEECs in terms of growth is much more likely if production and technology integration reinforce market integration between the existing EU and the CEECs. Otherwise, the CEECs could end up being integrated into the EU, but being isolated and marginalised in terms of production and technology linkages and excessively dependent on budgetary transfers. A proper understanding of the conditions for "deep integration" demands a better understanding of supply-side phenomena and, in particular, of the extent, factors and nature of production and technology linkages between the existing EU and the CEECs.

The varied experience of the current EU member states suggest very different degrees of integration through industrial networks although their levels of institutional and policy integration are identical. For example, integration of Spain and Greece through foreign direct investments (FDI) differs sharply despite the identical levels of institutional integration into the EU policy and legal structures. Different degrees

* This chapter is prepared within the project "The emerging industrial architecture of the wider Europe: the co-evolution of political and economic structures", funded by the UK ESRC programme "One Europe or Several?"

and forms of integration of the CEECs through FDI and subcontracting with the EU economies have already different effects on their restructuring and growth. For example, high productivity improvements of Hungary are greatly driven by productivity increases in foreign investment enterprises (Huyna 1999b).

In this chapter we provide the basis for better understanding of these issues by surveying the research that has been done on the issue of industrial networks between central and eastern Europe (CEE) and the EU. We survey financial, trade, FDI and production network aspects of the European integration. Rationale for reviewing these aspects of micro- or industrial integration is their close interconnection. Like any qualitative change globalisation of the wider Europe is a multidimensional phenomenon, and its patterns and driving forces are not clearly visible when we look only at its individual dimensions.[1] The very nature of the world economy is the existence of close interconnections between FDI, trade, technology transfer, finance, and labour movements. Based on such an understanding of globalisation the analysis of European integration will have to tackle its different dimensions.

Financial, trade, production and technology integration are different aspects of the integration of the CEE into the wider European economy.[2] Financial and trade integration can be described as "shallow" forms of integration, production and technology integration as "deep" forms of integration. The criteria for this distinction are the degree of mutual "lock in" of two economies or areas. Production and technology links are more "sticky" than only financial and trade links i.e. the interaction and knowledge exchange is in principle more intensive and long-lasting in the latter case. At one end of this integration spectrum are portfolio investments and at the other knowledge exchange through intra-firm flows. Although analytically we may separate these forms of integration in the real world forms of "shallow" and "deep" integration are complementing each other. Without financial and trade integration we cannot expect that technology integration will develop.[3] Equally, financial and trade integration between two areas may not be sustainable in a long term without deepening of production and technology integration.

---

[1] In the 1970s Hymer (1972) pointed out that the multinational corporate system has three related roles: international capital movements, international capitalist production and international government. In order to understand changes we have to look at all three dimensions and explore how they are related. Similarly to Hymer, Michalet (1994: 13) points out that the various dimensions of the world economy are tightly interconnected. He argues that it is no longer feasible to develop separate analytical frameworks for trade, financial markets, international monetary movements and migration as if these belonged to distinct fields that are subject to separate theories. Neither can the working of the world economy be understood by simply adding up distinct pieces of knowledge.

[2] For an extensive discussion on interaction of different forms of integration see Radosevic (1999).

[3] By *technological integration* we understand the integration of domestic enterprises into the dynamic learning processes of MNCs, whereby they become involved as active contributors and recipients in the production of knowledge for generating technical change

In continuation we briefly outline the main issues of the wider European integration by surveying financial, trade, FDI, and non-equity forms of integration. We also review several industry and firm levels studies, which have illuminated some aspects of the industrial integration of the wider Europe. Conclusions bring several research and policy issues which could form the basis for the future analysis.

9.2   Financial Integration

Financial integration is an essential element of economic integration of CEE. One of crucial constraints for growth of the CEE enterprises and their access to the EU markets is the lack of finance. Without strong financial integration of the CEE economies into the EU sources of finance production integration is unlikely to develop. For the CEECs this has been a fairly recent form of integration, especially firm issued equity and bonds-type of finance. According to EBRD (1998) three-quarters of all private net inflows entered the region during 1996-98, i.e. in the last half of transition period. Foreign direct investments are making up an increasing share of these flows but they still amounted to only one-third of the private capital inflows in 1997 and are attracted primarily to leading transition economies. The distribution of private equity issues has remained restricted to central Europe. Portfolio flows into the government securities markets have characterised a number of countries at less advanced stages in transition (ex. Russian Romania) (EBRD 1998: 77, 84). This suggest that the further behind the country is in the transition process it integrates financially through shallow forms of financial integration, like loans or investments in government Treasury-bills. The most remarkable difference between the CEECs are the terms of external borrowing, which are far more favourable for the more advanced transition economies (ibid: 85). The more favourable terms facilitate integration of the central European firms in foreign accessing market which than further facilitate heir production and technology integration.

We already pointed that different forms of integration are mutually complementary. Hence, the effects of financial integration of CEE are not confined to the level of financial aggregates, i.e. at the level of "shallow" integration. Financial globalisation is also directly changing strategies of MNCs towards CEE as well as of CEE enterprises and in that way is influencing the patterns of "deep" integration (Radosevic 1999a, chapter 4.1.). Foreign acquisitions amount to about 50 % of FDI in CEE (UNCTAD 1997). Also, CEE enterprises increasingly rely on non-bank investors through bond issues and global depositary receipts (GDRs). These trends are driven by the opportunities offered by the liberalisation of markets for corporate assets and opportunities to raise capital in quasi-equity forms through convertible corporate bonds or depositary receipts, but also by underdeveloped capital markets in CEE.

In overall, the financial integration is an essential facilitator of industrial networking of the wider Europe. Unfortunately, the assumptions underlying research on technology flows and financial flows exclude each other from their respective analytical frameworks. The assumptions underlying the financial models are insufficient for the inclusion of innovation. On the other hand, innovation studies rarely try to integrate financial aspects in their frameworks. When it comes to the industrial integration of the wider Europe, where both data on financial and technology forms of integration are sparse, these general weaknesses are several times magnified.

9.3   Trade Integration

Trade integration between CEE and the EU has developed quite extensively. For example, from 1988 to 1995 central Europe accounted for half of the increase in EU imports of industrial manufactured products from emerging regions outside the OECD. (Zysman/Schwartz 1998). The trade structure has been significantly changed as well. For example, exports in 1996 were only 72 % similar to those in 1998 for the region as a whole (Eichengreen/Kohl 1998: 21)

The comparative analysis of structural changes in EU - CEE trade as summarised in Radosevic and Pavitt (1999) shows that:

- All CEE countries are relatively weak in the production of R&D-intensive and capital goods.

- Within this general pattern, there are major differences amongst CEECs in their overall trading performance, and in the degree to which they are improving their performance in scale- and R&D-intensive sectors, and in capital goods.

- There seems to be a significant difference between the "capacities" of the CEECs in terms of measured R&D and human capital, and outcomes in terms of technology structure and unit prices of exports after 1989.

The analysis based on highly disaggregated trade data (Radosevic/Hotopp 1999) suggests the existence of three patterns of catching up and accompanied learning processes at micro level:

- Strengthening of export patterns based on labour intensive industries, like clothing and footwear, in all CEECs;

- The emergence of technology intensive export in transport machinery and the emergence of exports of electronic and electric products, especially in Hungary and Czech Republic;

- The maintenance of the previous strong orientation of export in commodities, which still remains an important part of the export product spectrum, but only in Bulgaria represents the most substantial share of export.

Technology intensive exports have the highest share in Hungary followed by the Czech Republic, while their share in Romania and Bulgaria is the lowest. Poland represents an intermediate case. Smith (1999) also shows that the ten accession states can be divided into three separate structural-economic categories with respect to changes in trade structure and transition.

This multiplicity of learning patterns suggests that the modes of involvement of the CEE into the global economy do not proceed in a linear manner, i.e. along one mode of adjustment, but represent a combination of several patterns. This finding is confirmed by unsystematic but persuasive FDI evidence where we find a broad variety of factory types (see Radosevic 1997). How are these patterns explained?

They are the result of the two groups of factors. On the one hand this process is shaped by inherited domestic structural features which are in inherited specialisation and in concentration/dispersion of the trade product structure. These supply side factors are coupled with features of EU demand, which is manifested through different forms of subcontracting (clothing), FDI (cars and car parts) or market demand (commodities). This suggests that the emerging trade patterns of CEECs cannot be explained without taking into account their micro-basis and the strategies of foreign companies, which are strongly shaping FDI and subcontracting patterns in the CEE. ECE (1994) also finds a clear link between FDI and trade developments as well as the strong contribution of FDI to export. Eichengreen/Kohl (1998) develop a similar conclusion. They argue that changes in relative prices have not been important in explaining trade performance. Also, the Association Agreements are unlikely to have played a dominant role in shaping CEE trade. The most important factors in shaping trade patterns are FDI and outward processing traffic (OPT). Their conclusion is that if trade policy mattered it did so via its effects on OPT and FDI (ibid: 33). This is confirmed also by data on foreign investment enterprises, which are much more trade intensive when compared to domestic enterprises (Hunya 1999a).

The organisational aspects of trade are very important in understanding its patterns and determinants. When compared to traditional trade theories the common point of departure of alternative approaches is the recognition that upgrading occurs in products and activities that are "organisationally related through the lead firms in a global supply chain" (Gereffy 1999: 39). The overwhelming presence of FDI suggests that these issues can no longer be ignored. However, a relative decrease of intra-firm trade compared with trade based on different forms of extra-MNE relationships, suggests that the analytical focus must be broadened to include different forms of industrial networks, like alliances and subcontracting.

The increased importance of the organisational basis of trade complicates further our understanding of determinants of trade. The non-arms' length forms of trade are not amenable to trade theories type of explanations based on different factor endowments. Equally, the non-equity forms of trade are not amenable to theoretical perspectives based on transaction cost approaches, like the internalisation approach (Rugman, Caves) or its amalgamations into an eclectic paradigm (Dunning). In addition, we do not have an alternative theoretical perspective to provide a unified explanation of the variety of forms and determinants of global industrial and supply networks. Each of these theories sheds light on a specific aspect of trade and international networks. For example, in the case of international alliances transaction cost, or the eclectic or resource based perspective help to illuminate only one aspect of the networks.

9.4 Foreign Direct Investments

International production integration of CEE is relatively the most researched from the FDI perspective. The general conclusion on direct effects of FDI is quite positive. For example, Eichengreen/Kohl (1998: 39) conclude that "(t)he correlation between FDI flows and increase in unit values, shifts in factor intensity and the other measures of trade performance ... strongly suggest that FDI is integrating the more advanced economies into multinational production networks, shifting these economies towards R&D-intensive and human capital intensive products". Also, the productivity of foreign investment enterprises in CEEC is similar to Austrian levels (Hunya 1999a).

In continuation we briefly review several important dimensions of FDI in CEE:

- *Structure of FDI.* The FDI stock of the CEECs diverges little from the normal pattern depicted by gravity models (Brenton *et al.* 1999). Three-quarters of FDI originates from EU based MNEs, and especially from Germany. Except in the case of neighbouring countries, explanations based on home countries seem to be irrelevant (Hunya 1999b).

  The share of FDI as a percentage of GDP of central European countries is now similar to the EU at around 17 %. Among CEECs only Hungary shows a very high degree of foreign penetration comparable to that of Spain and Ireland (Hunya 1998). The main common branch with high foreign penetration in the four CEE countries is the manufacture of transport equipment, most notably of motor vehicles. The second branch is the manufacture of electrical machinery and equipment (ibid).

The penetration of FDI in CEECs depends on industry-specific features and on the characteristics of the privatisation policies. The main features as described by Hunya (1998) are:

- FDI in CEECs follows worldwide differences in the corporate integration of industries
  * Technology intensive electrical machinery and car production are the main target
  * Textiles and clothing and leather are less internationalised by FDI than other branches
  * Foreign investors penetrate activities with relatively stable domestic markets, e.g. the food, beverages and tobacco industries
- Branches with low foreign penetration worldwide may have high foreign presence due to proximity (e.g. construction material)
- Privatisation by sales matters for FDI

The foreign presence has been so far relatively small in branches with great structural difficulties, and oversized capacities, like the steel industry and petrochemicals.

While CEE is primarily a recipient of inward FDI the region is also becoming a source of some outward direct investment flows. It seems that these investments reflect the ability of local companies to build and rebuild foreign trade and investment links. If we exclude Russian companies, than the leading outward investors from central Europe are Croatian and Slovenian companies.[4]

- *Motives of FDI.* The areas of FDI in CEE reflect the interest of foreign investors in access to CEE markets (trade and services, partly industry), the low cost sourcing advantages of CEE (industry) as well as the areas of biggest investment stakes in non-tradable sectors (telecoms; energy). Surveys on FDI by Lankes/Venables (1996), Meyer (1998) and others suggest the predominance of market seeking FDI. Factor cost considerations, as a single motivating factor is secondary. Only jointly with attractive markets do lower factor costs attract inward FDI (Meyer 1998).

The initial investments are those that are domestic market oriented. This explains why horizontal types of FDI projects are entering CEECs relatively early (Lankes/Venables 1996). The progress in transformation will make more CEECs hosts to vertical FDI bringing them into EU and world production networks.

---

4 Out of 10 top outward investors from CEE 5 companies are from Croatia (Podravka Group, Atlantska Plovidba, Pliva Group) and Slovenia (Gorenje Group, Adria Airways). These are followed by Czech Republic (Motokov, Skoda Plzen), Hungary (MOL) and Slovakia (VSZ Kosice). The biggest outward investor from the region is Latvian Shipping Co ($339m). In terms of industry structure top investors are from transport (3), food, appliances, trade, pharmaceuticals, conglomerate, fuel and metallurgy. *Gazeta Wyborcza*, September 28, 1999 as cited in Kubielas, NICOM Bulletin, October 1999.

- *Modes of entry.* In the early stage of transition foreign investors concentrated on joint ventures (J-Vs) with state-owned enterprises (SOEs) where they had minority positions. In the second phase, majority FDI was preferred. For example, in Hungary 81 % of nominal capital in foreign investment enterprises was with a minority share. In 1991, only 34 % of foreign capital was in companies with a minority foreign share (Hunya 1996). Joint ventures are the least preferred option when compared to FDI. It is likely that the EU enlargement will encourage mergers & acquisitions, rather than joint ventures. The reason for this is that joint ventures are acceptable only as a device to mitigate risk by allowing easier access to local actors, and for acquiring local knowledge (Lankes/Venables 1996).

- *Types of FDI.* The type of FDI varies significantly according to the host country's progress in transition (Lankes/Venables 1996). Advanced economies in transition have a higher share of export oriented FDI projects, subsidiaries are more integrated into MNCs and are more likely to be wholly owned. Also, countries which are further in transition have relatively few projects that have been abandoned or postponed, relatively more export supply oriented projects, and relatively more fully foreign owned projects (ibid: 1997). However, there is no evidence of a smooth functional relationship between FDI levels and transition level & risk (Lankes/Venables 1996). The penetration of FDI is not a function of only country specific institutional features but of a larger set of factors, like industry structure, firm strategies and specific FDI policies.

- *The effects of FDI.* FDI in CEE are not yet big employment creators. For example, the net employment effect of FDI in Hungary is only 33,000 jobs (Hunya 1996). The primary effects of FDI are in increasing the productivity and efficiency of acquired companies (Hunya 1997) (Knell/Hunya 1999) (Zemplinerova 1998). Their relatively higher share of investments and in R&D compared to domestic enterprises falls into this picture (Hunya 1999a; Inzelt 1999; Farkas 1997).

The effects of FDI are still localised in acquired or newly erected plants.

The initial econometric evidence on FDI spillovers at industry level in CEE is negative (Knell 1999).[5] The "enclave syndrome" has already become acute in the Hungarian economy and will probably follow in other countries as the size of foreign investments grows. Closer integration into the domestic economy will emerge as an important concern for CEE governments. However, the indirect effects of FDI should be seen within a time dimension. In many cases, the first investment phase (assembly activities) has been followed by further integration of local activities, adding the elements of procurement and logistics, supply and invoicing, product engineering, process and product development.

---

5 This is not surprising given the general conclusions of the literature on spillovers which suggest that there is not comprehensive evidence on the exact nature of magnitude of these effects. While earlier studies suggest that the effects are generally positive the increasing international division of labour within MNCs complicates the analysis (Blomström/Kokko 1998: 247).

As a rule, FDI are into branches that have relatively stable and promising or growing domestic markets. They are not made in collapsing branches with shrinking domestic market. In this respect their effects on industry structure may not be substantial.

Foreign investment enterprises are the main agents that deepen and extend the trade of CEE economies. They do not necessary deepen value-added. The effects of this on balance of payments are not always positive and, in the early phases in Hungary, they are a net burden on trade balance (Hunya 1996).

The effects of FDI on third countries are not clear. Increased FDI in Spain and Portugal in the late 1980s did not reduce investment flows to other EU countries. Brenton *et al.* (1998) show that FDI inflows to CEECs (1990s) did not have a clear negative impact upon the FDI of Spain and Portugal.

9.5   Non-Equity Production Networks

FDI are an important, but not the only, form of micro-level of integration between European "East" and "West". Alliances, and among them different forms of sub-contracting, are very often more important forms of integration than FDI. Given the high concentration of FDI in a few CEE countries (Poland, Hungary and Czech Republic) and on several sectors (cars, telecom, food, and trade) the overall impact of FDI on growth, with the exception of Hungary, seems slightly exaggerated. However, FDI is easier to monitor statistically and identify through business surveys or the business press. This also explains the stronger focus on them than on non-equity linkages. Contrary to FDI, alliances are far less defined; they are more heterogeneous and thus more difficult to compare. Also, the lack of systematic data opens up the danger of too hasty generalisation on a small sample. In continuation we discuss some of the research on alliances, subcontracting and Outward Processing Traffic (OPT) networks.

Alliances

The FDI data obscure extra-MNCs or non-equity networks which are probably even more important in CEE than FDI based links. Research by Schmidt (1998), and Dalago *et al.* (1997) considers also non-equity links, i.e. subcontracting and alliances. Their conclusions are that:

- Shallow modes of production co-operation, like arms' length transactions and contract work, dominate;
- CEE firms are integrated as sub-contractors in vertically structured networks conducted by "Western" firms;
- Low commitment of foreign firms;

- Local firms are involved in labour intensive stages which generate low value-added but are also experiencing quality, delivery and co-ordination problems.

Their research gives the impression that the "East" - "West" production networks are primarily contract manufacturing based at very low value added levels.

In Radosevic (1999b) we reviewed 26 alliances from Poland and the Czech Republic. Some of the main findings based on this sample are relevant to this discussion:

(1)   As a rule, firms grow either through foreign acquisition or networking (alliances), or through generic expansion, but one that relies heavily on networking. Generic expansion as a single strategy is rare.

(2)   Types of alliances or co-operative agreements in central Europe if judged on the basis of our sample are heterogeneous. Nevertheless, most alliances can be grouped into production alliances, or marketing alliances (in the software sector). The most widespread type of agreement is subcontracting.

(3)   Linkages generated by alliances are of both types, vertical and horizontal. However, vertical alliances were more common in our sample. This is to be expected given the frequent presence of subcontracting links. This also suggests that the alliances in central Europe are being driven more by unexploited market opportunities and cost differentials than by the wish to displace competition. A larger number of horizontal alliances would indicate a stronger presence of market share considerations. This seems to be much more of an issue in the case of FDI. Sadowski (1997), based on 4 digit SIC data concludes that the majority of M&A in central Europe have been horizontal acquisitions. This would suggest that in central Europe alliances are more prone to vertical relationships while FDI are more prone to horizontal links than alliances.

(4)   The comparison of cases suggests that the balance between generic expansion, alliances (networks), and mergers & acquisitions as modes of growth, reflects differences in firms' abilities to control technology, access to market and finance. If the enterprise is able to exert control over two of these three elements, it may ensure growth through alliances by trading them for the third, missing or weak element. However, the final outcome does not seem to be a direct function of the ability of enterprises to control access to technology, market and finance. It is not possible to fully understand alliances by only addressing their internal characteristics; how these mesh with their institutional context and with sectoral structural features must also be taken into account. The types and dynamics of alliances also reflect the political and legal situation of a country (privatisation, attitude towards FDI) as well as specific sectoral features in terms of technology, finance and markets. The profile of alliances is shaped through the interaction between firm-specific factors and capabilities, and sector- and country-specific factors.

Subcontracting

Subcontracting is considered here to be a specific form of alliance. In the simplest terms, subcontracting is contracting that partly contributes to the execution of a major contract (Nishiguchi 1994: 3). The firm awarding the contract to a subcontractor is the prime contractor, also termed the original equipment manufacturer (OEM), assembler, customer, purchaser, top firm, principal employer, or primary manufacturer. Although considered in statistics as ordinary trade, this seems currently to be the most important channel for technology transfer to CEECs. Subcontracting involves a variety of relationships like contract manufacturing; OEM subcontracting; cost- and speciality-based subcontracting, etc. This diversity of types of subcontracting makes statistical analysis of this type of production network difficult. However, the importance of subcontracting in sectors such as metal products, machinery, and car parts in countries like Hungary and the Czech Republic, and its growing weight in most of the CEECs, calls for a better understanding of this intra-industry type of trade in production processes. Subcontracting relationships enable the subcontracted enterprise to concentrate on production engineering, leaving marketing, and a proportion of the financial management, to the prime manufacturer. This is why for many CEE firms it is the only option if they are to survive and grow.

However, the problems involved in climbing the value-added ladder through subcontracting are not trivial. From a sample of 90 Hungarian subcontractors, Szalavetz (1997) showed that close co-operation with foreign partners brings considerable productivity improvements. In her sample, all processing firms received a transfer of technology or equipment and half the firms benefited from investment or working capital finance provided by the foreign partner. However, after the initial push the learning process gradually slowed and finally stopped completely. Szalavetz (1997: 5) points out that "Once the Hungarian company had undergone sufficient restructuring to ensure that co-operation can go smoothly, foreign partners abandon any further developmental effort". This occurred even in those cases where foreign partners had decided to increase their equity in the Hungarian company (ibid: 53).

This example illustrates the difficulties involved in deepening production integration and points to the discontinuous character of technological integration and the emerging structural barriers for CEE firms after initial productivity improvements.

Outward processing traffic

Outward processing traffic (OPT) is a specific form of subcontracting. OPT takes place when the principal firm supplies the host firm with intermediate products for processing which it then returns to the principal, which then markets the final output. This type of trade permits a relatively close relationship between the firms, and allows for quality control and detailed instructions to the host firm if necessary. OPT is a type of subcontracting which is widespread in every region and goes under different names. In the US it is called "production sharing" or the 807/9802 programme after the US trade tariff numbers which allow for this type of trade. In UK it is known as "jobbing contract". EU imports after OPT from extra-EU countries reached ECU12.0bn in 1994, with CEECs making up 47.8 % of the total. OPT trade is highly industry-specific and, in some sectors, constitutes the main form of trade (Naujoks/Schmidt 1994). Very low unit value of import after OPT from the CEECs indicates that the bulk of this trade is in labour-intensive sectors and is based on labour cost differentials, indicating a low level of technological integration (Sdogati 1996).

However, in general, wages are not the only factor that explains the spread of OPT subcontracting. Based on Asian OPT networks, Gereffi (1999) suggests that the determinants might be more numerous, and include wage exchange rates, trade policies (quotas, preferential tariffs) and social and cultural factors, such as ethnic ties, common language, and common historical legacy. Eichengreen/Kohl (1998) show that OPT has not promoted the movement into higher value added exports in the case of Visegrad 4 (Hungary, Poland, Czech and Slovak Republics) but that a case can be made for it in the second tier of CEEC economies (Bulgaria, Romania). In contrast, Pellegrin (1999) is much more optimistic regarding the restructuring potential of OPT subcontracting. She argues that "(t)here is not evidence of the destruction of local export capabilities resulting from the cessation of OPT activities due to wage increases. It is rather the opposite trend which is observed: OPT activities go hand in hand with the strengthening of trade performance" (p. 11). She also notes that these outcomes are mostly due to the relocation activities of German medium-sized firms "which illustrates the important role the latter have in bringing about such transformations" (Pellegrin 1999: 19).

This brief overview of the wider European integration through non-equity production networks gives some very relevant but also very limited, and sometimes contradictory, insights. As a result, generalisations and hypotheses run ahead of empirically based conclusions. Given very limited research in this area this is not surprising.

9.6   Industry Studies

Patterns of global industrial networks are, to a great extent, industry specific. Foreign equity investments are in car assembly, mixture of equity and subcontracting in car parts industry, subcontracting in clothing, co-operative agreements in software - these are typical sectoral modes of entry in CEE. They do not reflect CEE specificity alone but also the technological features of the sectors. Research on systems of innovation and international political economy shows that "there are striking variations across sectors in the nature and kind of authority and how much it, or they, intervene with the play of market forces. Compared with differences between national laws and institutions affecting the economy, the differences are apt to be much greater between sectors of the world market economy" (Strange 1996: 187). However, as pointed out by Hall (1997) the principal problem is to specify precisely which variables in sectoral relationships (firm, national, sectoral, international) have the most impact on corporate strategy and economic outcomes.

The radical change in the industrial structure of individual sectors led to changes in supply and demand and to a complete change in the position of enterprises in CEE, which is very much sector-specific. For example, CEE telecom equipment producers have developed from being producers of outdated switching equipment to becoming dependent subsidiaries localising state-of-the-art technologies (Mueller 1999). Computer producers had to completely abandon the idea of producing their own mini-computers, and were transformed into PC assemblers (Bitzer 1998). New software firms have become customisers of generic solutions in close co-operation with foreign software providers (Bitzer 1998b). Car complexes of the former socialist period have been transformed into networks led by foreign assemblers and reorganised with the help of first-tier foreign suppliers (Richet/Bourassa 1999). Domestic car part producers have become subcontractors serving foreign-controlled assemblers (Havas 1999a, 1999b). In his analysis of CEE electronics, Linden (1998) concludes that Hungary and the Czech Republic "are connected, in varying but increasing degrees, to international electronics production networks. Foreign investment is the primary vehicle for this integration, and Hungary has moved the furthest along this path, positioning itself as a major low-cost supply base in the region. Czech electronics firms, remaining independent, do not have the same degree of production linkages with foreign firms, despite a strong underlying skill base in precision engineering. Poland has been able to leverage its larger market to some extent, but has been slower than Hungary to attract export oriented investment" (p. 27).

As a result of the "deverticalisation" of the socialist production networks, which went hand in hand with the opening of domestic markets and foreign investments, the local value- added has been drastically reduced. However, competitiveness and productivity have improved dramatically, especially in enterprises benefiting from foreign investment (Hunya 1998, 1999a).

Studies on the car industry, computers and software, telecommunications, shipbuilding and the food industries by Bourassa/Richet (1998), Bitzer (1998a, b), Mueller (1998), Bitzer/von Hirschhausen (1998) and Charpiot-Michaud (1998), and their comparisons in Radosevic (1999a), suggest that market demand is essential for the restructuring process. In those sectors, or subsectors, where domestic demand is growing, progress in industrial modernisation is more likely. However, demand alone is not sufficient for restructuring to occur; rising demand could be satisfied through imports. Sectoral studies suggest that the pace of this process is also likely to be determined by gaps in technology and finance. If both the finance and technology gaps are small, as in the case of PC assembly, customised software and, to a certain extent, in the food processing industry, restructuring can be expected to take place. If, on the other hand, technology and/or finance gaps are problematic, difficulties in modernisation, or more significant country differences, for instance in telecommunications services or car assembly, are likely.

Sectoral differences arise from inherited similar technological levels and institutional deficiencies, primarily in capital markets. The sectoral features produce common technology, market, and financial control features, and in addition the patterns of industrial networks are not a direct function of these factors. The sectoral factors seem to operate only as tendencies and not as determining factors. The final pattern is determined by firm-specific factors, which involve not only domestic but also foreign firms, and also by broader political and institutional factors strongly reflected in privatisation.

Industry studies also show that the growing demand for products or services does not automatically generate demand for domestic S&T or for S&T links. For Central and Eastern European S&T systems, the growth of demand for S&T is essential. The break-up of large firms in CEE has reduced demand for innovation to levels lower than product demand would suggest. Also, the proliferation of new small firms probably generates a different type of demand for R&D and innovation, to which domestic R&D cannot respond immediately. This may partly explain why, that despite the recovery in the CEECs, we do not discern the emergence of dynamic sectoral innovation systems. In addition, innovation systems almost everywhere are "hybrid" systems, embodying complex public/private interdependencies. This suggests that, even where there is a critical mass of demand for domestic innovation and technology, a plethora of other missing factors may be related to the hybrid character of systems of innovation that may prevent its emergence.

In the most restructured sectors in CEE, emerging structural barriers to further industrial upgrading are becoming evident (Dyker 1999). In these sectors, CEECs may reach the limits of industrial upgrading based only on foreign direct investment or foreign-led modernisation, both of which are characterised by intra-firm productivity improvements in foreign investment enterprises but not yet by increasing foreign-domestic innovation linkages.[6] Based on the analysis of the CEE car industry Dorr/Kessel (1999) conclude that "(t)he limited spillovers have undoubtedly helped widen the gap which has arisen between international and local companies. It is currently impossible to see how far this deficit is simply an effect emanating from

---

6 For an overview of evidence on this see Radosevic (1999a).

the starting phase, which will decline over time. This division of industry into two very unequally developed areas, which has occurred, is seen critically from the perspective of the national economies. It resulted, however, not just from the group strategies but also from the very weak industrial and structural political governance of the transformation countries" (ibid: 23). However, the majority of CEECs is still struggling to integrate into international production networks, and for them integration at any technological level is seen as a solution.

The sectoral differences in modes and degree of industry integration suggest that their overall effects on growth and industry upgrading can be understood better by analysing sectoral rather than aggregate effects and factors. For example, the evidence offered by van Tulder (1998) on the car industry, suggests that "(t)he competitive balance in the European car industry, has not yet been much affected by the creation of cross border production networks (p. 46). However, "CEE FDI in car industry has contributed to additional overcapacity in European car industry." (ibid.). If this is the effect in the sector where FDI are very significant, in CEE the overall effects of "East" - "West" industry networks on growth and industry in EU may be still limited. However, this conclusion may be completely wrong. We should be aware that we are chasing a moving target about which we know relatively little. For example, business analysts expect in the next few years a wave of foreign investments in CEE steel and chemicals, sectors that have been so far largely bypassed. Also, the nature of the existing FDI and networks is changing. The specificity of foreign-led industrial networks in CEE is the multiplicity of their technological levels. For example, the issue that stands out clearly from sectoral studies on the car industry is the dichotomy between CEE operating as *a low cost base* by hosting the lower end of the value-added chain and CEE operating largely as *a complementary production base*. Van Tulder (1998) finds both patterns operating in the CEE car industry while Dorr/Kessel (1999), based on an analysis of VW/Audi investments in CEE, conclude that the dominant strategies in the car industry were not in the pattern of "the extended workbench, but companies with structures and products, firms which could prove themselves as manufacturing operations with complicated technology and also with suitably qualified people" (p. 21). This multiplicity of technological positions of CEE operations of EU producers suggests that the overall effects are not easily discernible and are very much sector specific.

9.7 Firm Studies

Sectoral level studies show the diversity of the international industrial networks that are emerging in CEE. This diversity runs not only across sectors but also within sectors and is strongly shaped by the individual strategies of foreign investors. For example, van Tulder (1998) shows that the shape of international production networks in the European car industry largely runs along the lines of four strategic groupings: frontrunners (Volkswagen, General Motors, Fiat and Renault), followers (PSA, Ford), peripheral (Suzuki, Daewoo) and (voluntary) lock out (BMW, Toyota, Nissan, Daimler Benz) networks. Van Tulder/Ruigrok (1998) conclusion is that "(e)ach group of firms share different strategic intentions for the region. Followers and lockout networks largely see the region as a still limited market. Peripheral firms primarily use the region as an entry into the Western-European market. Front-runner firms adopted the most sophisticated (and also most difficult to manage) strategy: they aim at the region as a production site for cheap re-imports back into the home base, they see it as a source for lower-end world cars and components, and they see the region as a market" (van Tulder/Ruigrok 1998: 46). This points to the role of individual firms in shaping patterns of industrial networks. Industrial networks, which individual firms are part of, have a significant impact on the nature of success of the strategies that firms pursue. Equally, individual firms are able to shape the patterns of adjustment to a large number of firms with whom they are in co-operation or competition. The models of operations of foreign firms in CEE are diverse. As industry studies show, they range from operations where CEE functions as a low cost base to those where CEE operates largely as a complementary production base. In the upper range of business models we find establishment of new production models as in the case of VW/Skoda (Dorr/Kessel 1997) (Brezinski/Fluchter 1998), or integrated affiliates like in the case of GE/Tungsram and ABB (Barham/Heimer 1998). However, the most widespread seems to be the position where CEE enterprises operate as extended workbenches or localizers (Lankes/Venables 1996; Radosevic 1999a). The opportunities opened by the European integration lead to interesting new business models of CEE firms, which are based on extensive use of subcontracting and alliances. A very illuminating case in this respect is Hungarian Videoton (Szalavetz 1997).

9.8   Conclusions

The extent and nature of the linkages that emerge between the "East" and "West" of Europe will strongly shape the competitive dynamics and industrial development in CEE but also in the EU. The accession of the CEECs into the EU raises the issue of whether "East" - "West" industrial networks will be a factor in improving the growth prospects of the enlarged EU or whether they will deepen the differences in levels of development and undermine prospects for more balanced growth. Although market integration is a necessary objective of enlargement, it is in no way a sufficient condition of dynamically efficient outcomes for an enlarged EU. Convergence of the CEECs in terms of growth is much more likely if production and technology integration reinforce market integration between the existing EU and the CEECs. Otherwise, the CEECs could end up being integrated into the EU, but being isolated and marginalised in terms of production and technology linkages and excessively dependent on budgetary transfers. This survey of the extent and of the main issues involved in the CEE - EU industrial integration pointed to some facets of the emerging industrial architecture of the wider Europe.

First, financial, trade, production and technology aspects of integration of the CEE into the wider Europe are developing at different speed and intensities across different countries. Research suggest that the "shallow" forms of integration, i.e. trade and finance linkages are developing fast and extensive. Especially, trade integration between central European countries and the EU has developed tremendously. However, we cannot understand the driving forces of the EU – CEE trade integration without taking into accounts the organisational basis of trade and industrial networks which underpin trade integration.

Second, we now have a relatively better picture of the spread and driving factors behind FDI. However, our understanding of non-equity modes of industrial integration is still very sketchy. The available sectoral studies show important sectoral differences in the modes and extent of industrial integration. While existing studies have generated a mass of statistical data their value is diminished by their limited comparability.

Third, the specificity of EU integration, when compared to other regional integrations in the world, is the strong policy and institutional integration. The integration process is top-driven and aims at "deep" institutional integration. However, the viability of political and institutional integration of the wider Europe cannot be separated from the breadth and depth of integration at firm-level. Policy (macro) and production (micro) integrations are driven by different forces and have certain degrees of autonomy, i.e. they can be developed independently of each other. However, in the long-term they should be compatible and reinforce each other. If the disparity between depth and breadth of micro/production and macro/institutional integration becomes too great this will increase both economic and political costs for the EU and the accession countries and could undermine the enlargement raising diverse security concerns. The need to ensure compatibility between policy and production integration and thus reduce the social costs of their incompatibility raises a whole set of new policy and management issues.

9.9   References

BARHAM, K./HEIMER, C. (1998): *ABB. The Dancing Giant. Creating the Globally Connected Corporation.* London: FT. Pitman Publishing

BITZER, J./VON HIRSCHHAUSEN, C. (1998): *Shipbuilding: Final Report - 'Industrial restructuring'.* TSER project 'Restructuring and reintegration of S&T systems in economies in transition'. Berlin: German Institute for Economic Research.

BITZER, J. (1998): *Computers: Final Report - 'Industrial restructuring'*. TSER project 'Restructuring and reintegration of S&T systems in economies in transition'. Berlin: German Institute for Economic Research.

BITZER, J. (1998b): *Software: Final Report - 'Industrial restructuring'*. TSER project 'Restructuring and reintegration of S&T systems in economies in transition'. Berlin: German Institute for Economic Research.

BLOMSTRÖM, M./KOKKO, A. (1998): Multinational corporations and spillovers, *Journal of Economic Surveys*, 12, pp. 247-277.

BRENTON, P.F./MAURO, D./LUCKE, M. (1998): Economic Integration and FDI: An Empirical Analysis of Foreign Investment in the EU and in central and Eastern Europe, *Kiel Working Paper* No. 809. Kiel: The Kiel Institute of World Economics.

BREZINSKI, H./FLUCHTER, G. (1998): *The Impact of Foreign Investors on Enterprise Restructuring in the Czech Republic - Findings based on a case study of Skoda Volkswagen*. Paper presented at the First International Workshop on transition and Enterprise Restructuring in Eastern Europe, Copenhagen Business School.

CAVES, R. (1996): Multinational Enterprise and Economic Analysis, *Cambridge Survey of Economic Literature*, Cambridge University Press.

CHARPIOT-MICHAUD, F. (1998): *Restructuring the food processing industry in Eastern Europe for international competition: economic analysis with special reference to the role of S&T policy in East and West, Final report - 'Industrial restructuring'*. TSER project 'Restructuring and reintegration of S&T systems in economies in transition'. Berlin: German Institute for Economic Research.

DORR, G./KESSEL, T. (1999): Restructuring Via Internationalization. The Auto Industry's Direct Investment Projects in Eastern Central Europe. *WZB Discussion Paper FS II 99-201*. Berlin: WZB.

DORR, G./KESSEL, T. (1997): *Co-operation and Transfer. Difficulties of Mutual Understanding in Czech Joint ventures and the Case of Skoda - Volkswagen*. Berlin: WZB.

DUNNING, H.J. (1991): The Eclectic paradigm of international trade, PITELIS, N./SUGDEN, J. (Eds.) *The Nature of the Transnational Firm*. London: Routledge.

DUNNING, H.J. (1993): *Multinational Enterprise and the Global Economy*. Reading: Adison Wesley.

DUNNING, H.J. (1997): *Alliance Capitalism in Global Business*. London: Routledge.

DYKER, D. (1999): Foreign Direct Investment in the Former Communist World: A Key Vehicle for Technological Upgrading, *Innovation*, 12, pp. 345-352.

EBRD (1999): Transition report 1999. Ten years of transition. London: European Bank for Reconstruction and Development.

EBRD (1998): Transition Report 1998. London: European Bank for Reconstruction and Development.

EICHENGREEN, B./KOHL, R. (1998): The External Sector and Development in Eastern Europe, *Journal of International Relations and Development*, I (1-2), pp. 20-45.

ELLINGSTADT, M. (1997): The Maquiladora syndrome: Central European prospects, *Europe-Asia Studies*, 4, pp. 7-21.

ECE (1994): *The Impact of Foreign Direct Investment on the Trade of Countries in Transition.* Results of a Preliminary Survey. Geneva: UN Economic Commission for Europe.

FARKAS, P. (1997): *The Effects of Foreign Direct Investment on Research, Development and Innovation in Hungary*, Working Paper No. 81. Budapest: Institute of World Economy.

GEREFFI, G. (1999): International trade and industrial upgrading in the apparel commodity chain, *Journal of International Economics*, 48, pp. 337-70.

GUERRIERI, P. (1999): 'Technology and structural change in trade of the CEE countries', DYKER, D./RADOSEVIC, S. (Eds.) *Innovation and Structural Change in Post-Socialism: A Quantitative Approach.* Dordrecht: Kluwer Academic Publishers.

GUERRIERI, P. (2001): Technology, structural change and trade patterns of eastern Europe, HUTSCHENREITER, G./KNELL, M./RADOSEVIC, S. (Eds.) *Restructuring of Systems of Innovation in Eastern Europe.* Aldershot: Edward Elgar. (forthcoming)

HALL, A.P. (1997): *The Political Economy of Europe in an Era of Interdependence.* Cambridge MA: Center for European Studies, Harvard University, mimeo.

HAVAS, A. (1999a): *Changing patterns of Inter- and Intra- Regional Division of Labour: Central Europe's Long and Winding Road.* Budapest: National Board for Technical Development, mimeo.

HAVAS, A. (1999b): *Local, Regional and Global Production Networks: Restructuring of the Automotive Industry in Hungary.* Paper produced within the TSER project Restructuring and Re-Integration of S&T systems in Economies in transition, ROSES, Paris.

HUNYA, G. (1996): Foreign direct investment in transition countries in 1995, *Vienna Institute Monthly Report*, 1, pp. 2-8.

HUNYA, G. (1997): Large privatisation, restructuring and foreign direct investment, SALVATORE, Z. (Ed.) *Lessons from the economic transition: central and eastern Europe in the 1990s.* Dordrecht: Kluwer Academic Publishers, pp. 275-300.

HUNYA, G. (1998): *Relationship between FDI, Privatisation and Structural Changes in CEECs.* Paper prepared for the 'Conference on Privatisation, Corporate Governance and the Emergence of Markets in Central - Eastern Europe'. Vienna: The Vienna Institute for International Economic Studies.

HUNYA, G. (1999a): *The role of FDI in industrial restructuring in CEECs.* Paper presented at the SSEES Conference on FDI in Poland. School of Slavonic and East European Studies – University College London. Vienna: The Vienna Institute for International Economic Studies, mimeo.

HUNYA, G. (Ed.) (1999b): *European Integration through FDI: Making Central European Industries Competitive.* Cheltenham: Edward Elgar.

HYMER, S.A. (1972): 'The Internationalization of Capital', Colloques Internationaux du CNRS, *The Growth of the Large Multinational Corporation*, No. 549, Rennes.

INZELT, A. (1999): The Transformation Role of FDI in R&D: Analysis based on Material from a Databank, DYKER, D./RADOSEVIC. S. (Eds.) *Innovation and Structural Change in Post-Socialist Countries: A Quantitative Approach.* Dordrecht: Kluwer Academic Publishers.

KNELL, M. (1999): Foreign investment enterprises and productivity convergence in central Europe, HUNYA, G. (Ed.) *European Integration through FDI: Making Central European Industries Competitive.* Cheltenham: Edward Elgar.

KNELL, M./HUNYA, G. (1999): National Innovation systems and foreign affiliates in eastern Europe, KNELL, M./HUTSCHENREITER, G./RADOSEVIC, S. (Eds.) *Restructuring of Innovation Systems in Central Europe and Russia.* Cheltenham: Edward Elgar.

KRYKOV, V./MORE, A. (1996): *The New Russian Corporatism?* A Case Study of Gazprom, Post-Soviet Business Forum. London: The Royal Institute of International Affairs.

KUBIELAS, S. (1999): Transformation of technology patterns of trade in the CEE countries, DYKER, D./RADOSEVIC, S. (Eds.) *Innovation and Structural Change in Post-Socialist Countries: A Quantitative Approach.* Dordrecht: Kluwer Academic Publishers.

LANDESMANN, M. (1997): Emerging patterns of European Industrial Specialization: Implications for Labour Market Dynamics in Eastern and Western Europe, *Research Reports*, No. 230, The Vienna for Comparative Economic Studies (WIIW).

LANDESMANN, M. (1999): The Shape of the New Europe: Vertical Product Differentiation, Wage and Productivity Hierarchies, KNELL, M./HUTSCHENREITER, G./RADOSEVIC, S. (Eds.) *Restructuring of Innovation Systems in Central Europe and Russia.* Cheltenham: Edward Elgar.

LANKES, H.P./VENABLES, A.J. (1996): Foreign direct investment in economic transition: the changing pattern of investments, *Economics of Transition,* 4, pp. 331-47.

LINDEN, G. (1998): *Building Production Networks in Central Europe: The Case of the Electronics Industry.* BRIE, Working paper 126, http://socrates.berkeley.edu/~briewww/pubs/wp/wp126.html.

MARKUSEN J.R./VENABLES, A.J. (1997): Foreign direct investment as a catalyst for industrial development, *NBER Working paper series 62411*, http://www.nber.org.

MEYER, K. (1998): *Direct Investment in Economies in Transition*. Cheltenham: Edward Elgar.

MICHALET, C.-A. (1994): Transnational corporations and the changing international economic system, *Transnational Corporations*, 3, pp. 9-21.

MUELLER, J. (1998): *Restructuring of the Telecommunications sector in the West and the East and the Role of S&T, Final Report - 'Industrial restructuring'*. TSER project 'Restructuring and reintegration of S&T systems in economies in transition. Berlin: Berlin School of Economics (FHW).

NAUJOKS, P./SCHMIDT, K.D. (1994): *Outward Processing in Central and Eastern European Transition Countries: Issues and Results from German Statistics*. Kiel Working Paper No. 631. Kiel: The Kiel Institute of World Economics.

OECD (1998): *The Competitiveness of Transition Economies*. OECD, WIFO and WIIW, Paris.

PELLEGRIN, J. (1997): *International Business and the European Integration Process: the Example of Outward Processing Traffic between the EU and the Central and Eastern European Countries*. DPhil thesis, Florence: European University Institute.

PELLEGRIN, J. (1999): *German Production Networks in Central/Eastern Europe. Between Dependency and Globalisation*. WZB, Discussion paper FS I 99 - 3304. Berlin: WZB.

RADOSEVIC, S. (1999a): *International Technology Transfer and Catch-Up in Economic Development*. Cheltenham: Edward Elgar.

RADOSEVIC, S. (1999b): *Restructuring and reintegration of S&T systems in economies in transition, Final Report*. TSER project. Brighton: SPRU.

RADOSEVIC, S. (1997): Technology Transfer in Global Competition: The case of Economies in Transition, DYKER, D.A. (Ed.) *Technology of Transition*. Budapest: Central European University Press.

RICHET, X./BOURASSA, F. (1998): *Restructuring of the East European Car Industry. Final Report - 'Industrial restructuring'*. TSER project 'Restructuring and reintegration of S&T systems in economies in transition, ROSES, Universite Paris I.

RUIGROK, W./VAN TULDER, R. (1995): *The Logic of International Restructuring*. London: Routledge.

SCHMIDT, K.-D. (1998): *Emerging East - West Collaborative Networks: An Appraisal*. Kiel Working Paper No. 882. Kiel: The Kiel Institute of World Economics.

SCHMITTER, C.P. (1990): Sectors in Modern Capitalism: Modes of Governance and Variations in Performance, BRUNETTA, R./DELL'ARINGA, C. (Eds.) *Labour Relations and Economic Performance*, Proceedings of a conference held by the International Economic Association in Venice, Italy. London: Macmillan.

SMITH, A. (1999): *The Competitiveness of East European Exports in EU Markets.* Paper presented a the SSEES Conference on 'Poland and the Transformation in Europe: The Role of Foreign Direct Investment'. London: UCL.

STRANGE, S. (1996): *The retreat of the state. The diffusion of power in the world economy.* Cambridge: Cambridge University Press.

SZALAVETZ, A. (1997): *Sailing before the wind of globalization. Corporate Restructuring in Hungary.* Working Paper No. 78. Budapest: Institute for World Economy.

TOTH, G.L. (1994): Technological Change, Multinational Entry and Re-Structuring: The Hungarian Telecommunications Equipment Industry, *Economic Systems*, 18, pp. 179-198.

UNCTAD (1994) World Investment Report. Geneva: UNCTAD.

VAN TULDER, R./WINIFRIED, R. (1998): *European Cross-National Production Networks in the Auto Industry: Eastern Europe as the Low End of European Car Complex.* BRIE Working Paper 121, May, http://brie.berkeley.edu/~briewww/pubs/wp/wp121.html.

10. East German Industrial Research: Improved Competitiveness through Innovative Networks

Franz Pleschak, Frank Stummer

10.1 Formulation of the Problem

Although the number of industrial R&D staff has been consolidated and the competitiveness and profitability of external industrial research institutions improved in East Germany, obvious differences between the features of East and West German innovation activities still exist; in particular those of R&D. Innovation activities in East Germany are characterised by a high share of small and medium-sized companies which often show a lack of potential for complex R&D projects. Both the share of industrial R&D staff and the industrial share of economic activity are lower than in West Germany. Moreover, less funds are available for R&D staff (Meyer-Krahmer *et al.* 1998; Pleschak *et al.* 2000).

One way of reducing the problems emerging from this situation is the synergy of innovative potential in networks. This is not only necessary because of the complex starting position of East German industrial research, but also in order to meet the demands on industrial research which arise from international innovation competition. This contribution shows the chances for innovation activity which evolve from the establishment of networks, and the conditions which are required for their successful use.

10.2 Requirements of Innovation Activity

Innovation activity is characterised by *complexity* in many different respects. Firstly, it affects the innovation process which reaches from the idea for new products, and solutions to processing and organisational questions, through research and development, production planning, and market introduction up to broad market penetration. These are problem solution processes characterised by risk, the division of labour, numerous interactions with surrounding fields, and inter-disciplinary co-operation. The generation of new knowledge requires close co-operation between all participants, in particular regarding application-oriented basic research and the industrial application of R&D results from different scientific disciplines.

Moreover, the complexity of new technology is increasing. This demands not only higher standards of knowledge, skills, and experience on the side of the R&D staff, but also forces all innovation actors to communicate intensively. Co-operation is indispensable if one wants to meet the technological challenge of international competition.

Decisions about innovation activity are also very complex in character. In the case of an institution, they require reflection and a method of working which overlaps and links individual departments; moreover, customers, suppliers, marketing partners, investors and other actors from this area must be integrated into the preparation and the finding of decisions. Especially market conditions and customers' requirements set economic, technical, and time standards for the whole innovation process.

Within the innovation system, the division of labour is being intensified. In East Germany, this is shown by *external industrial research institutions*, which have come into existence since 1990 due to the new profile of the research field; their economic consolidation has been realised in the past years due to their closeness to the market and to customers. The emergence of specialised R&D institutions working in close co-operation with the market does not represent a specifically East German situation but is the general tendency of innovation systems. It is understood that such R&D institutions must complement each other and become integral systems again, either by co-operating in networks, or through direct co-operation contracts. In many large companies, de-centralisation and outsourcing lead to a relaxation of traditional forms of organisation, and create the chance for independent economic activities in specialised departments. Several factors recommend a deeper *division of labour between innovation actors*:

- Due to their economic situation, small companies cannot afford to finance specialised R&D staff, the utilisation of which would be too limited.

- In order to enlarge their capacity, to reinforce their innovative capacity, or to introduce results of basic or applied research into their innovation process, medium-sized and large companies make temporary use of additional potential.

- Growing complexity and the system character of products and processes require special knowledge within the R&D sector, as well as experience in many fields of natural science and technology, which are not available to an individual company.

- R&D requires special instruments and equipment, processes, information, etc., the utilisation of which does not meet with the requirements of operational efficiency, and which can often be provided by external firms faster and more cost-effectively.

- On the one hand, specialisation generates advantages due to the gaining of competence, synergy effects, the division of risks, and the completion of capacities. On the other hand, specialisation also stimulates learning effects on the side of partners, and closer relationships between research and the industrial utilisation of research results.

- The tendency towards concentration on core businesses leads to the outsourcing of those fields of activity which companies themselves cannot fulfil efficiently.

The deeper division of labour within the innovation process forces partners into *increased co-operation*. This is particularly true for the co-operation between small and medium-sized companies, as well as with external industrial research institutions. Sustained competitiveness is only guaranteed for those companies or R&D institutions which are able to adapt, and which are willing and ready to co-operate. The value of an innovation is not determined by a single innovator but by the whole innovation system in the figurative sense of a close weave of relationships between the innovation-relevant actors. Especially in East Germany with its industrial culture of small enterprises, a high level of co-operation and inter-links is required to enable innovative developments. Large companies – especially those with their own R&D departments – are almost non-existent in East Germany.

In networks as a special feature of co-operation, complementary resources, several actors, and inter-dependent co-operative activities are tied up to reach the common objective "innovation process" in the framework of *co-operation*, which must *not necessarily be formally regulated*. Synergies between the networking actors reduce their risks and accelerate the R&D process. The staff's qualification is supported by the knowledge and technology transfer taking place in the network. Networks can provide favourable framework conditions for the control of innovation requirements such as complexity, inter-disciplinary co-operation, and comprehensiveness (Koschatzky/Zenker 1999).

The fact that the East German situation is characterised by torn networks, broken-off business relationships, a lack of time and money for the establishment of new business relationships, potential network partners' lacking confidence in the efficiency and survival of East German companies, hinders innovation (Gemünden/Ritter 1999). In East Germany there is a lack of large companies serving as the core of crystallisation for networks; co-operation abilities, as well as fundamental information about potential partners, must be improved for regional actors.

For this reason external industrial research institutions serve the function of a "hinge" between producing companies and basic or applied research. They contribute to knowledge and technology transfer; in innovative networks, their role as a service company close to production is primarily the support of small and medium-sized companies. If this co-operation is not realised, then companies renounce the chance to integrate results of basic and applied industrial research into their new products and processes, and thus the character of a unique position for their innovation. Thus, overcoming the non-interlocked structure of East German companies is a central task of innovation management (Gemünden *et al.* 1997).

An essential element of the innovation system in East Germany is formed by the external industrial research institutions which have come into being, and which act on the market as R&D service companies and institutions. They work close to the market and are thus forced to consider industrial problems as the point of departure of their own R&D. Due to their R&D potential they are able to combine basic industrial research and the practical utilisation of R&D results. They open up innovation potential and make it useful for small and medium-sized enterprises situated close together, and are effective not only on an international but also on a regional level.

10.3   Joint Operation of Innovative Potential in Networks

10.3.1   Network Actors

As has been shown, the small and medium-sized innovative companies or R&D institutions in East Germany do not have the economic force to meet the requirements of international innovation competition. There is a lack of specialists in the different fields of technology and scientific disciplines, which prevents them from coping with the necessary inter-disciplinary character of R&D, as well as with the system character of innovations. Close inter-linking with basic research represents a problem due to a lack of funds. Small and medium-sized enterprises also encounter problems for the internationalisation of innovation process. However, in order to have access to top level research and to open up new markets, internationality is indispensable.

Moreover, it is hard for small enterprises to take advantage of effective labour division, to finance efficient instruments and equipment, and to achieve cost advantages through higher output quantities. Nevertheless, if they want to be competitive, even SMEs must meet with these requirements.

To overcome this problematic situation, one option could be the collaboration of the following *innovation actors* in networks:

- Companies with a common interest in R&D, production and marketing, which, through co-operation, would be enabled to offer and market system solutions to their customers;

- Universities and non-university research institutions which, on the one hand, are interested in the utilisation of results from their basic and applied research, and which, on the other hand, need feedback about experience with practical use, as well as problem recognition on the side of industrial users;

- Sales partners which, due to their knowledge of target markets, customers' wishes and requirements, as well as the market participants' conventions, contribute to the opening-up of markets through network partners, which ease traditional market structures, and impart marketing know-how while establishing their own position on the market;

- Suppliers which find partners from the network actors for steady relationships between customers and suppliers, and which contribute to reinforcing the network through innovative developments (equipment, materials);

- Customers which have a reference effect as key customers, which first benefit from the advantages of a product and thus contribute to the distribution on the market;

- Innovative service companies, which either introduce special R&D know-how or which, as consultants, help to introduce operational methods, to carry out market surveys, or to improve efficiency through process analyses;

- Public agencies which contribute to the regional integration of networks, and which derive impulses from these for regional development;

- Competitors, which play an efficient role as partners for the enforcement of regulation and admission activities, as well as for the elaboration of norms and standards, or which improve their market chances as a bidders' association.

Consequently, networks are based on the links between the networking partners, the economic objectives they have agreed upon, and their common interests. As was proved, innovations not only depend on teachable and learnable knowledge which can be logically derived, or reproduced in a structured way, but also on knowledge emerging from an environment based on experience (empirical or tacit knowledge), which comes into existence more in an intuitive and situative way, and which is therefore not generally accessible. Personal relationships between all network partners represent an important source of such knowledge and its practice-effective utilisation. Innovations based on this kind of knowledge instead of generally accessible knowledge are hard to imitate (Noppeney 1997). Due to the knowledge and technology transfer taking place in the network, staff qualification is supported, and learning processes are generated.

Spatial aspects of networks play an important role for the development of East German innovation potential: the establishment of networks is most advanced in some centres of crystallisation, e.g.

- Jena (bio-technology, optics and laser industry),

- Dresden (micro-electronics),

- Berlin and its surroundings (medical technology, information and telecommunication technology),

- Chemnitz (mechanical engineering),

- Freiberg (coating technology).

In order to initiate regional co-operation between all participants of the innovation process, the Federal Ministry of Education and Research has organised the promotional competition "InnoRegio" in the new Federal States. It is aimed at contributing to the development of new products, processes, and services, and to the establishment of a congruent regional profile on an educational, research, and economic level; moreover, all actors should be involved in the regional innovation dialogue. In this way, the reinforcement of the innovation competence and competitiveness of specific regions is planned. Networks are an important pre-condition for this.

The fact that far from all innovation actors co-operate in networks is shown by a study of the eight technology centres (TGZ) of the free state of Thuringia (cf. Table 10-1).

Table 10-1: **Frequency of contacts between the technology centre of Thuringia and other technology actors (n=8 centres)**

Type of contact	Technology Actors				
	University	Technical college	Non-university research institution	External research institution	Companies
Exchange of information	6	7	5	4	6
Joint projects	2	2	2	2	6
Joint events	6	5	6	2	8
Supply of consultants	6	6	3	3	7
Common use of technical equipment	4	3	2	2	7
Transfer of research results	5	2	4	1	4

Source: Gross/Pleschak (1999: 39).

Table 10-1 shows the frequency and type of contacts between these technology centres and other technology actors. The university and remaining scope of knowledge, the economic structure and innovation potential form the framework conditions for co-operation between regional actors. Networks provide the best pre-conditions for the establishment and development of contacts.

10.3.2   Advantages of Networks

Innovative networks possess the advantages shown in Table 10-2. The regional concentration of inter-linked companies can lead to considerable productivity advantages, better chances for innovation, and more favourable pre-conditions for founding activities (Porter 1999).

Table 10-2:   Advantages of working in innovative networks

• Gaining know-how- and competence, in particular during transfer to new technology, acquisition of external knowledge, learning effects
• Synergy effects in the areas of R&D, marketing, manufacturing, and information
• Improved market situation due to system solutions
• Double work is avoided, capacities are complemented
• Risk sharing and risk reducing
• Utilisation of advantages of size and specialisation
• Reduced development periods, acceleration of market introduction, advantages of flexibility
• Development of long-term business relationships
• Reduced transaction costs for the opening up of co-operation relationships
• Stronger position in negotiations with large companies and regional groups with a common interest
• Co-ordination of actions regarding the network environment (capitalists, consultants)
• Benchmarking of the internal efficiency

The efficiency and competitiveness of network partners are increased, the chances for success of business relationships are improved due to joint R&D projects, common strategies for industrial property rights and questions of certification, customer and market related activities, production-related tasks, marketing activities and relationships with suppliers. Due to the transfer of experience and its feedback for the company, learning processes are started. In this way, every participant derives advantages which otherwise are only typical for large companies. The degree to which management is actually improved depends on the structure, content and intensity of relationships in the network, as well as on the respect of the principle of quid pro quo by all partners.

10.3.3 Co-operation in Networks

Above all, radical innovations form the basis of long-term economic growth. Therefore, companies and R&D institutions must carry out innovation activities in close co-operation, based on common objectives. This procedure is necessary to assure, for example:

- that the point of departure for research is the real task of industrial utilisation, in order to ensure that research results can be realised on a technical level;
- that the absorptive capacity for preliminary research is available in the companies, and that their own development of new principles, effects, or processes will be passed on to new products;
- that production technical requirements are taken into account during construction, and that conclusions are drawn from constructive design in time for manufacturing investments;
- that R&D is confronted with customers' and market requirements, as well as with economic criteria.

This type of collaboration reaches far beyond general network relationships and requires formally regulated co-operation, based on objectives already agreed upon by the partners. However, networks are the basic unit for such co-operative relationships. Project-related co-operation brings specialised potential together with specific strong points; thus, access to external knowledge is improved, development periods are shortened, the risk of each participating company is reduced, and innovation projects which a single company could not realise due to a lack of financial, personnel and material resources are made possible. It is understood that co-operation is also problematic: strategies must be revealed, independence is reduced, and there is additional expenditure for co-ordination and communication. Therefore, co-operation management is needed in order to decide when, with which partners, how, and for what objective co-operation should be initiated (Pleschak/Sabisch 1996).

In the framework of jointly carried out innovation projects, co-operation between actors should work in such a way that only minimal expenditure is necessary for co-ordination at the interfaces, and that the loss of time and information is as limited as possible. Although it is possible to reduce the number of interfaces by the integration of functions within the innovation process, there are economic limits to such reductions due to the actual co-operative character of innovation processes and the advantages of specialisation arising from them. An important point of departure for the control of interfaces are strategies commonly decided on by all involved partners. These take shape in the framework of networks.

Although co-operation is indispensable for innovation success, it is not unproblematic (cf., for instance, the contribution of Fritsch in this volume). This is shown by several facts:

- Due to the uncertainty and the risk of innovation processes, a clear contractual regulation of co-operation is impossible. Contracts always include "soft" elements.

- In general the user is not able to assess the value of a R&D result which is transferred from research to development.

- The spreading of research results to other users often cannot be avoided, so that sole exploitation is endangered.

- R&D-related information is often locally anchored, and its transfer is linked with considerable expenditure. The understanding of information is based on the whole supply of information, which is naturally not available for the partner; thus, his ability to understand is limited. In many cases, information can only be documented in a limited way, and close personal contacts are required for comprehensive understanding.

- The user of R&D results himself does not always have the absorptive capacity to recognise the potential use of the R&D performance, to adapt this external knowledge to his own purposes, and to further develop them.

- The ability to communicate and to learn is not sufficient on the side of the partners in order to correctly grasp and reflect on problem situations in the innovation process; there could also be a lack of consciousness or confidence regarding co-operation with partners, or the efficiency of potential partners is called in question.

Conditions for successful co-operation can be improved by the learning and decision-making processes which take place in networks.

East German innovative potential does not yet make full use of the possibilities of R&D co-operation (Pleschak *et al.* 2000). Table 10-3 shows the frequency of R&D co-operation between East German innovative companies or institutions and other innovative actors. However, the type of co-operation is not specified. It can be anything from an information exchange to co-operation regulated by contract. It is shown

- that research institutions co-operate mostly with other university and non-university research institutions,

- that R&D service companies have more R&D co-operation than producing innovative companies, except for co-operation with universities. In fact, in the area of

R&D projects, producing companies co-operate more with university research institutions than R&D service companies,

- that only a minor role in R&D co-operation is played by those innovation-supportive service companies which do not have their own R&D available such as consultants and transfer institutions.

Table 10-3: **Frequency of R&D co-operation between companies and institutions and other innovative actors** (in %)

Innovative actors	Producing companies (n=59)	R&D service companies (n=49)	External research-institutions (n=34)	Non-university research – institutions (n=14)
Producing companies	54	69	76	71
Service companies with their own R&D	20	51	44	21
Innovation supportive service companies without their own R&D	12	20	32	7
Non-university research institutions	49	57	88	86
Universities	73	53	76	86

Source: Pleschak *et al.* (2000: 88).

There is a relatively limited frequency of co-operation between companies and R&D service companies on the one hand, and non-university research on the other hand. It is true that companies and R&D service companies have their own considerable R&D potential; however, an extension of R&D co-operation with research institutions certainly also includes attractive options to improve efficiency and competitiveness.

The problem is even more obviously shown by a study of the areas in which R&D co-operation takes place. According to Table 10-4, only one fifth of the companies or R&D service companies have co-operation relationships with basic research; on the other hand, only every third external research institution has connections with basic research. There are obviously more co-operation relationships in the area of applied research; however, the share is only about 60 % for companies and R&D service companies. Most of their co-operation subjects are product innovation, with a distinctly lower share of process innovation.

Table 10-4: **Frequency of areas of R&D co-operation** (in %)

Areas of R&D co-operation	Producing sector (n=59)	R&D-service companies (n=49)	External research-institutions (n=34)	Non-university research in-stitutions (n=14)
Basic research	17	16	35	79
Applied research	61	57	91	71
Product development	85	90	79	43
Process development	41	45	47	63

Source: Pleschak *et al.* (2000: 90).

In order to provide industry with better options for R&D co-operation, the Ministry of Economy and Technology has launched the promotional programme PRO INNO (innovation competence for middle-class companies), which fosters the following projects:

- Joint co-operation projects between companies,

- Joint co-operation projects between industry and research institutions,

- Co-operation projects which are transferred from a company with a R&D mission to a research institution,

- Initial industrial projects,

- Staff exchange between companies, or between companies and research institutions,

- With the help of this promotion, technical and economic risks of R&D as well as the transaction costs companies have to bear in the case of co-operation should be reduced; in this way, industry should be encouraged to make more efforts in favour of R&D.

10.3.4   Pre-conditions for Networks

Networks function provided that every participant introduces competence into the network, that there is an absorptive capacity for external knowledge, as well as the capacity to adapt and further develop the knowledge. Network partners must be able to communicate and to learn in order to be prepared for problems and for the partners. Consequently, networks are characterised by a culture of interactions. Co-operation is based on personal contacts and the respect of non-written rules. Typical hallmarks of successful networks are openness, honesty, confidence, the common definition of objectives, readiness to co-operate, as well as consensus about and co-ordination of interests. Contractual regulations should counter the risks of uncontrolled drifting away of knowledge and revealing of strategies.

A decisive pre-condition for innovation success is a *competent management of network relationships*. In general, every network actor has his own forms of co-ordination, which are seldom coherent with those of his partners. Therefore, networks require independent forms of steering co-ordination, as well as harmoniously suitable instruments of co-ordination (commissions, plans, communication meetings, etc.) (Munser 1998). Table 10-5 summarises the pre-requisites for successful performance of innovative networks. In order to counter the risk of unilateral individual benefits from network performance, the presence of neutral network managers can make sense.

Table 10-5: Pre-requisites for the successful performance of innovative networks

• Learning and communicating ability of the network partners
• Ability to absorb and deliberate upon information
• Concentration on the partners' problem situation, similar problem solution behaviour, comparable competence for the solution of problems
• Long-term, stable business relationships, processes of interaction, and bonds.
• Independent profile of every network partner, which is complementary to that of the other network partners
• Openness towards and confidence in network partners
• Avoiding hierarchies in the network
• Redundancies in the network to avoid dependence
• Voluntary co-operation
• Connection of resources and utilisation of the co-operation
• Spatial proximity between of the partners without delimitation from supra-regional knowledge and information transfer
• Economic advantages due to co-operation
• Development of innovation-relevant relationships with the surroundings
• Interface management and moderation

In order to avoid the loss of time and information, as well as a lack of co-ordination, network co-operation requires an *interface management*. In networks very different actors co-operate, with equally different interests and objectives. On the one hand, it is obvious that the individual participant has improved prospects for success due to the integration into a network, with the resulting system effects and lower transaction costs; on the other hand, individual participants need a learning process in order to integrate the chances offered through networks into their own work; moreover, co-ordination is required for the assessment and generalisation of experience and to reach agreement upon common strategies.

The tasks of a *network management include* (Gemünden/Ritter 1999):

- Preparing the ground for business relationships,
- Information exchange between the different companies,
- Network marketing and sale of network components,
- Planning, organisation and financing of joint projects,
- Planning, co-ordination and joining of common activities,
- Establishment and maintenance of the infrastructure,
- Co-operation with regional decision-makers,
- Creation of an image for the network,
- conflict management.

Networks are a form of collaboration between actors involved in an innovation process which can lead to synergy-producing effects. Due to the fact that information and the knowledge about innovation always come into existence, and are made use of, on a local level, networks are often characterised by a strongly local dimension. Their functioning depends on the level of regional innovation potential, on the congruence of the elements of the innovation system with regard to the contents (according to branches, R&D areas, degrees of specialisation), and on the level of inter-connection between the actors. Regions provide the framework for people with the same cultural background and understanding of problems (Koschatzky/Zenker 1999). Spatial proximity in a region enables the kind of personal contacts which are an important pre-requisite for learning in networks. Since the elements of networks, as well as their objectives and methods, are characterised by regional conditions, there is no universally valid scheme for the establishment of networks.

10.4 Final Remarks

Despite the fact that networks have already been established in East Germany and that there is openness and readiness for participation in networks, the co-operation between the elements of the innovation system does not yet meet the requirements. However, interconnection is crucial for the functioning of an innovation system, and particularly important for increasing the efficiency of the East German innovation system, which is characterised by many small and medium-sized companies. To realise the above mentioned effects, the establishment and development of networks should be supported for a limited time; however, the establishment of networks as an end in itself should be avoided. Favourable effects for the co-operation of innovative potential in networks are:

- Support of the network management in order to relieve the involved partners in their short-term additional charge of establishing networks, and to reduce their workload of moderation and co-ordination. At the beginning, synergies in the network are not opened up automatically but are essentially the result of learning processes. Provided that there is sufficient experience on the side of management, networks become autonomous due to their economic advantage and the network partners' own interest (knock-on promotion).

- The support of innovation projects which target the network as a whole, and which cannot be financed solely through market relationships due to their high risk.

- The support of networks as a whole where they play a particularly important role for regional development, where they give impulses for regional growth, or development chances for structurally weak regions.

- The support of innovation projects which require work in close co-operation in order to train co-operation and the division of labour (joint projects for radical innovation with a high degree of novelty, high risk, and high complexity). This would help to reduce the capacity and financing bottlenecks of small and medium-sized companies, as well as of research institutions, in the starting phase of comprehensive projects; moreover, the bringing together of potential of basic research, applied research, development, manufacturing and market introduction would be guaranteed.

10.5   References

FRITSCH, M./MEYER-KRAHMER, F./PLESCHAK, F. (1998) (Eds.): *Innovationen in Ostdeutschland - Potentiale und Probleme.* Heidelberg: Physica-Verlag.

GEMÜNDEN, H.G./RITTER, TH. (1999): Innovationserfolg durch technologieorientierte Geschäftsbeziehungen – Ein Vergleich zwischen Ost- und Westdeutschland, TINTELNOT, C./MEISSNER, D./STEINMEIER, I. (Eds.) *Innovationsmanagement.* Berlin: Springer.

GEMÜNDEN, H.G./RITTER, TH./RYSSEL, R./STOCKMEYER, B. (1997): *Innovationskooperationen und Innovationserfolg – Empirische Untersuchungen unter besonderer Berücksichtigung der Unterschiede zwischen Ost- und Westdeutschland.* Studie für das BMBF, Bonn: Bundesministerium für Bildung, Wissenschaft, Forschung und Technologie.

GROSS, B./Pleschak, F.: (1999): *Technologie- und Gründerzentren im Freistaat Thüringen – Untersuchungen zur Leistungsfähigkeit.* Stuttgart: Fraunhofer IRB.

KOSCHATZKY, K./ZENKER, A. (1999): *Innovative Regionen in Ostdeutschland – Merkmale, Defizite, Potentiale.* Karlsruhe: Fraunhofer ISI (Arbeitspapier Regionalforschung Nr. 17).

MUNSER, R.K. (1998): Die Koordination kooperativer Forschung und Entwicklung, *io management.* Zürich.

NOPPENEY, C. (1997): Quellen der Innovation, *io management.* Zürich.

PLESCHAK, F./FRITSCH, M./STUMMER, F. (2000): *Industrieforschung in den neuen Bundesländern.* Heidelberg: Physica-Verlag.

PLESCHAK, F./SABISCH, H. (1996): *Innovationsmanagement.* Stuttgart: Schäffer-Poeschel.

PORTER, M.E. (1999): Unternehmen können von regionaler Vernetzung profitieren, *Business manager.*

11. Innovation Networks and Regional Venture Capital Companies in Germany – Experiences for Central and Eastern European Countries

Marianne Kulicke

032 038
624 632

11.1 Introduction

Since the beginning of the 1990s, venture capital financing has gained significance as an instrument of innovation policy used by the Federal Government and the Federal States of Germany. As a parallel development, partly stimulated by government measures, the private venture capital market has shown itself to be very dynamic since 1997 (cf. Lessat *et al.* 1999; Leopold/Frommann 1998.). Following a period of stagnation with new investments of barely DM 1 billion annually, a total new business amount of approximately DM 3.8 billion was invested in 1998. According to the figures for the first six months of 1999 reported by the non-profit making German Venture Capital Association, another high growth rate of DM 4.5 to 6 billion is expected for 1999.

At present, young and small companies with a high potential for growth have no difficulties in finding a large number of venture capitalists to provide not only capital but also managerial support for establishing a business.

The financing situation is distinctly less favourable for the great number of technology firms which offer less chances for capital gain for investors, and nevertheless have high risks in financing and establishment. Besides private investors or other companies with limited requirements regarding the rate of return, only promotion-oriented or less profit-oriented venture capital companies come into consideration for these technology firms. In general, the business activity of both groups is limited to a specific region such as a German Federal State or, as is shown by the venture capital subsidiaries of savings banks, to the region of interest of their parent companies. In the following contribution, these will be called "regional venture capital companies".

Due to the fact that equity capital earnings do not cover the high costs of venture capital management, regional venture capital companies are not able to provide extensive assistance for the establishment of young companies from their own resources. It is therefore all the more important for them to be part of an innovation network in which they fulfil the financing function, whereas other networking partners are in charge of managerial assistance or consultation. Managerial support refers to the manifold tasks which are linked to the establishment of a new company or to the development of a small firm, such as the orientation of products and services towards market requirements, the establishment of productive staff, the development of reliable contacts with customers and suppliers, the opening-up of foreign markets, etc.

Regional venture capital companies play an important role on the German venture capital market. Among other projects, they also finance foundations and high-tech companies with a low growth potential, and in this way close an equity supply gap for those areas which are important on a national economic level but are too unattractive for private equity for reasons of profitability.

Such supply gaps also exist in the Central and Eastern European transition countries. It is true that every industrial or transition country should primarily aim at the financing of companies in general, as well as at innovation, through private (venture) capital; however, at the present stage of the re-structuring process in these countries, private venture capital companies, i.e. which are purely gainfully active, will not be able to fulfil this function in the medium term.[1] On the one hand, there are not many such capitalists with experience in providing capital for the whole field of possible reasons for financing. On the other hand, there is a lack of sufficient demand on both a level of quantity and in particular of quality, from companies with a high potential for profit and growth; these companies could represent profitable capital investment projects for private investors. Many companies still show a high level of market uncertainty; some of them still have outdated production units, poor quality standards and limited knowledge of international marketing. Consequently, they are less attractive for national and international venture capitalists.

Nevertheless, for companies in the re-structuring process as well as for newly founded technology companies, such financiers could be ideal partners who would be able to link capital provision to differentiated management support in order to compensate for internal deficits and to assist with (new) strategic orientation as well as the realisation of innovation strategies. However, promotion-oriented or less profit-oriented capitalists can only afford assistance in commercial and marketing

---

[1] Reference is made to the situation of the Czech venture capital market as an example; cf. Bross/Walter (1998).

questions where the regionally available innovation-supportive infrastructure – i.e. an innovation network – is integrated with it.

11.2   Transaction Costs and the Venture Capital Market

The venture capital market is an example of a market where transaction costs, which influence market participants' activities, play an important role. The following types of costs can be differentiated:

- *Information costs* are spent on the search for and evaluation of appropriate partners. This applies both to companies requiring capital, and to investors (venture capital companies).

- *Contractual costs* which result from the negotiation of investment conditions and the formulation of contracts.

- *Controlling costs* are not only generated through the control of reliability and contractual fidelity (mostly by investors), but also through management support, generally needed by the portfolio companies in order to reach investors' objectives linked to the capital supply ("value added").

The level of these transaction costs depends for example on the number of potential contracting partners, on the distribution of information about the contracted project, on the degree of uncertainty regarding the further development of technological, competitive, societal or legal framework conditions, on the frequency of transactions, and, finally, on the spatial distance between the contracting partners. The features of these determinants define the level of the "transactional barrier" between the capital supply side and the capital demand side.

As well as exit and failure costs, information, contracting and controlling costs reduce investors' capital gain. In order to cover the above-mentioned costs through capital gain and to make an attractive profit for their investors, those venture capital companies which are exclusively gainfully active have very high expectations regarding the profit potential for their investee companies (portfolio companies). Promotion-oriented venture capital companies have to *minimise* or *externalise* these costs: their chances of making a profit are limited because their task is to promote the economy. The same is true for the middle group of less profit-oriented venture capital companies, in particular the sector of savings banks.

The above-mentioned transaction costs have different levels of importance for the gainfully active and for the non-profit making capitalists: promotion-oriented venture capital companies mostly enter silent partnerships, which are repaid by the investee company after the contracted maturity date. Therefore, the costs of concluding the contract and exiting are distinctly below those of gainfully active venture capitalists. In the case of venture capital failures, most of the costs are generally covered by public promotional programmes or guarantees. Most of the transaction costs are accounted for by the procurement of information regarding the choice of appropriate portfolio companies, as well as for the controlling, information and consulting functions during the life span of a venture capital investment. Limited personnel capacities on both a level of quality and of quantity force promotion-oriented venture capitalists to externalise as many costs as possible for the selection and supervision of their investments. This is mostly assured by a (regional) supportive network.

Regarding profit-oriented venture capital companies, which primarily secure direct investments in their portfolio companies, the above-mentioned types of transaction costs (information, contracting, controlling, exit and failure costs) are distinctly higher than for the other types of venture capital companies, which mostly enter silent partnerships. The search for and selection of promising investment projects is more complex, the negotiation of investment conditions and the set of agreements are less standardised than in the case of silent partnerships; moreover, the different desinvestment types (e.g. trade sale, initial public offering) imply a considerable work-load for the investor. However, the most important cost burden generally results from the management support linked to the investment (in view of "value added") and from the coverage of failures from the portfolio stock.

11.3   The Role of Regional Venture Capital Companies in Germany

After many years of stagnation, the number of new venture capital companies in Germany has increased considerably: national investors have founded new venture capital companies as well as other capital investment companies, international companies have established German branches, individual Federal States have initiated innovation funds, and a multitude of savings banks have founded venture capital subsidiaries. This founding activity is also reflected in the precipitous rise in the number of members of the BVK[2] (Bundesverband deutscher Kapitalbeteiligungsgesellschaften - German Venture Capital Association e.V.), which shot up from 86 at the end of 1996 to 122 in October 1999. The actual increase in the number of active venture capital companies is probably even higher, due to the fact that only a small part of the total number of barely 70 venture capital companies of savings banks and their holding organisations are BVK members.

---

[2]   Source: The respective BVK almanacs.

During the past years, the financing instrument "venture capital" has distinctly gained importance in Germany for the financing of young and small technology companies. Approximately 35 %[3] of the DM 3.84 billion venture capital which was invested in companies in 1998 for the first time, went to high tech branches (DM 1.34 billion). In comparison with the preceding years, an obvious growth rate can be seen: in 1995, the quota was only 21.1 %, then 26.6 % in 1996, and only 25.4 % in 1997.

The German venture capital market is dominated by three groups of venture capital companies (cf. Figure 11-1):[4]

❶ Definitely profit-oriented venture capital companies which only invest in companies with a high potential for growth or profit (venture capital and corporate capital companies, venture capital companies of large and private banks, of insurance companies, etc.);

❷ companies (primarily Middle Class Venture Capital Companies[5] and funds launched by the Federal States for innovation financing), initiated by the individual Federal States, which are aimed at the promotion of the economy;

❸ venture capital companies established by institutes of the savings banks' organisation, people's banks, and rural credit co-operatives, etc. In spite of their orientation towards trade and industry, the investment of venture capital is part of their mission to support the regional economy.

The activity of clearly profit-oriented venture capital companies is not restricted to specific regions; instead, they invest on a nation-wide level or enter into commitments in several German Federal States. The major part of the newly invested venture capital volume per year accounts for them, and most of the members of the BVK belong to these.[6]

Since the beginning of the 1970s, promotion-oriented venture capital companies have represented an instrument of structural policy in order to strengthen the equity base of new and middle-class companies. They also played an important role at the beginning of the 1990s, when venture capital was included in the promotional pol-

---

3   According to the distinction of different high tech sectors made by EVCA (EVCA 1999).

4   A detailed illustration of the individual groups of venture capital companies is found in Lessat *et al.* (1999: 119ff.).

5   "Mittelständische Beteiligungsgesellschaften (MB)" (Middle Class Venture Capital Companies) as a self-help institution of the economy, are restricted to specific Federal States. Their objective is to supply equity for as many of SMEs as possible in their region of interest, in order to strengthen regional economic power and the level of qualified jobs offered. Due to the limited level of their own resources, they primarily have to fall back on public promotional offers.

6   Only one third of the venture capitalists organised in the Association of German Venture Capital Companies BVK are restricted to a specific region or an individual Federal State (cf. illustration of the BVK members in the web).

icy of the Federation and of the States as a new instrument of innovation promotion for young and small technology companies. If one takes as a basis the number of investments instead of the investment volume, a considerable part of investments account for such promotion-oriented investors: according to an estimate, a little less than two thirds of the total of about 3,800 business investments held on the German market at the end of 1998 account for such promotion-oriented investors.[7]

Figure 11-1: **Regional orientation and profit orientation of investors on the German venture capital market**

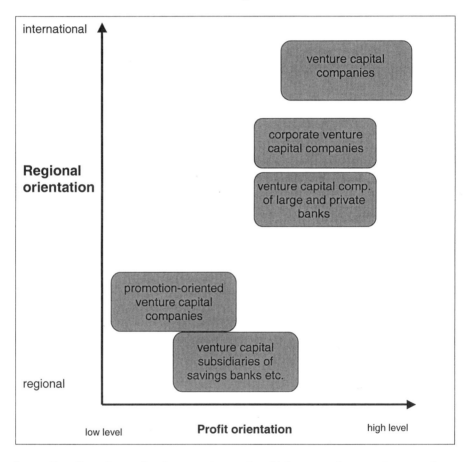

Innovation financing only plays a minor role which cannot be exactly quantified due to a lack of statistical data. The economic structure of the respective regions is

---

7 Estimate based on the number of investee companies indicated by the BVK members, in propor-tion to the number of investments. It is taken into account herewith that the number of invest-ments includes both initial and follow-up investments.

reflected by the composition of investee companies according to branches. During the past years, promotion-oriented venture capital companies have shown a higher degree of involvement in innovation financing of young and small enterprises. The non-financial support of their portfolio companies is generally restricted to financial and commercial questions; because of their role as silent partners, and due to the large number of investee companies and their personnel capacity, only limited support can be provided.

Most Federal States have initiated special promotion-oriented innovation venture capital companies in the past years,[8] these are supposed to enter into silent partnerships with young and small companies and to provide more extensive consultation than middle class venture capital companies do. In fact, the latter are limited regarding support in marketing and sales, or assistance with international product marketing. The establishment of innovation venture capital companies was aimed at closing the technology financing gap for those companies where the growth potential is too limited for profit-oriented venture capitalists, but whose capital requirements, risk structure and need of support surpasses the abilities of middle-class venture capital companies.

Venture capital companies of institutes from the savings banks' organisation, people's banks and rural credit co-operatives, etc., invest in growing small and medium-sized and partly young companies, and generally provide limited management support (primarily regarding commercial and financial questions). They mostly enter silent partnerships and sometimes also direct investments; their primary interest is to make a profit through current repayments. Due to their regional limitation they have distinctly lower expectations regarding the growth potential of the companies needing financing, and their services include a wider range of funding opportunities since otherwise the potential of appropriate companies from the region would be too limited.

It is true that there are more venture capital companies from savings banks than promotion-oriented venture capital companies; however, their funding volume and their regions of interest are limited. Consequently, they participate in distinctly less companies than promotion-oriented venture capital companies. Older companies mostly have 15 to 25 portfolio companies.[9]

---

8 E.g. Bayern Kapital Risikokapitalbeteiligungs GmbH – Bavarian Capital Venture Capital Company Ltd.

9 In comparison, the number of investees companies of Middle Class Venture Capital Companies (MBG) by the end of 1998 was respectively: MBG Baden-Württemberg (982), MBG Saxony (194), MBMV Mittelständische Beteiligungsgesellschaft Mecklenburg-Vorpommern (90), BayBG Bayerische Beteiligungsgesellschaft/ Bavarian MBG (465).

Due to the sharp increase in the number of new profit-oriented venture capital companies, corporate capital companies and venture capital companies of large and private banks as well as of insurance companies, etc., all of which are active on a supra-regional or even international level, the relative importance of regional venture capital companies in Germany is presently being reduced. Even though promotion-oriented venture capital companies are subject to considerable restrictions (cf. Kulicke 1998: 18ff.), they still play an important role:

- They facilitate the start-up financing until the company has reached a state of development which makes them attractive for other investors.

- They act as co-financiers in large-scale projects which a sole investor would not be able to finance alone.

- Where founders have the option of covering part of their funding requirements through a silent partnership, their need to integrate a direct partner and to cede company shares is reduced.

- Limited venture capital repayments reduce the financing costs and thus increase the profit of shareholders, including that of gainfully active venture capitalists; consequently, the company becomes more attractive to them.

However, they are part of a (public) supportive network, which they make use of in order to assist their portfolio companies.

11.4   Integration of Regional Venture Capital Companies into Innovation Networks

Regional venture capital companies are elements of different networks which overlap to some extent. According to their functions, these are:

- a network for the identification and evaluation of potential investee companies (portfolio companies),

- a network to secure global financing (start-up financing, follow-on financing); (in this case it is a financing network rather than an innovation network), and

- a network for assisting the portfolio company during the operating time of the investment.

The partners acting within these networks are mostly financial institutions, public consultation and promotion agencies, private consultants (management consultants, tax advisors, qualified auditors), other venture capital companies, universities, and non-university research institutions.

For regional venture capital companies, integration into a network aimed at the support of young and small (technology) firms firstly offers the advantage of limiting expenditure for customer contacts. Secondly, if they manage to inform their network partners about their offer of risk-carrying capital and about their conditions of investment, if they take advantage of events, of information brochures, of the public relation activities of these partners for the presentation of their own venture capital policy and for their own publicity campaign, then they have the chance to become known as capital providers within their target group.

The importance of a network for the identification of appropriate portfolio companies is not only shown by the amount of the annual deal flow (the annual number of requests addressed to venture capital companies by companies in need of capital), but also by its quality. A good indicator of the deal flow quality is the investment rate (the rate of investments concluded of the total number of financing enquiries). A survey of 33 venture capital investors acting in the early stage financing of technology companies (Wupperfeld 1996) showed that German MBGs reach an investment rate of little less than 40 % on average. This was considerably higher than the rate realised by the profit-oriented venture capital companies acting on a supraregional level (in general well below 10 %).

The reason for the higher quality of capital inquirers is not only a higher degree of acceptance and public image of MBGs at potential portfolio companies, but also the chance they offer to fall back on efficient regional networks. Within these networks, a pre-selection of potential portfolio companies is primarily carried out by financial institutions. The pre-selection includes a review of the chances of realisation and of the credit worthiness of the company in need of capital. Synergy effects such as reduced costs for information and review occur for venture capitalists, due to the fact that the criteria of financial institutes regarding the granting of credits are the same as those promotion-oriented venture capital companies or venture capital companies of savings banks, people's banks, or rural co-operatives have for their silent partnerships.

Due to the co-operation with credit institutions, only the detailed analysis for the evaluation and selection of promising companies has to be carried out by the MBGs themselves, whereas the rough analysis as the first part of the two-stage venture capital examination is done by the (potential) lender. Savings banks co-operate in a similarly close way with their venture capital subsidiaries. In this way, regional venture capital companies are able to externalise part of their transaction costs.

However, neither credit institutions nor regional venture capital companies are able to evaluate the technology and marketing aspects of innovative projects of young and small technology companies. Therefore, both lenders and equity investors involve external institutions such as technical experts and consultants, to different degrees, and on either an informal or on a contractual basis. In this way, the MBG of Baden-Württemberg draws on expert opinions of the Steinbeis Foundation in order to evaluate innovation-bound venture capital investments to be funded by promotional programmes of the Federal State; the Steinbeis foundation has a compact network of transfer centres, which are essentially represented at the technical colleges of Baden-Württemberg.

Regarding regional venture capital companies' assistance of portfolio companies, not many empirical studies exist (e.g. Wupperfeld 1996; Lompe *et al.* 1998). It is emphasised in both descriptions by venture capitalists and reports on founders' experience that in the past, assistance was mostly limited to the rights linked to silent partnership, which are: information, control, and the right to decide. Regional venture capital companies essentially have an observation and controlling function;[10] in certain cases they arrange for contacts. Only few of them offer more intensive management support and therefore refer to the regionally existing innovation network (such as the S-UBG AG, venture capital company of the economic region of Aachen). Most regional venture capital companies cannot provide extensive management support but are active partners within a network of innovation-supportive institutions. Their relationships with young and small technology companies are generally reduced to their financing function and the observation of rights and obligations emerging from the respective form of capital supply. Consultation and support of portfolio companies are provided by other partners within the innovation network, with whom they co-operate on an informal basis.

Recently, regional venture capital companies have been involved in innovation networks to a higher degree. This is partly due to the competitions launched by the Federal Ministry of Education and Research (BMBF), "BioRegio"[11] and "EXIST - University-based start-ups",[12] which stimulate and encourage the establishment of regional innovation networks.[13] A reinforcement of such networks is also aspired to on a Federal States level, for example by the GO! programme in North-Rhine-Westphalia. In most of these regions, regional venture capital companies are among

---

[10] Regarding the different functions of venture capital companies, please refer to Wupperfeld (1996) or Kulicke/Wupperfeld (1996).

[11] On a nation-wide level, the three model regions of the BioRegio competition are: the BioRegion triangle between the Rhine and Neckar rivers around Heidelberg/Mannheim, the BioRegion of Munich, and the BioRegion Rheinland (surroundings of Köln, Jülich, Aachen). A special vote was attributed to the BioRegion of Jena.

[12] The programme "EXIST" supports initiatives in five model regions: "bizeps" – initiative for the support of foundations of the Bergisch-Märkische region of Wuppertal/Hagen, "dresden exists" in the region of Dresden, "GET UP" – generation of technology-oriented innovative foundings with high potential in the technology triangle between Ilmenau, Jena, and Schmalkalden, "KEIM" – Founding impulses for the extended technology region around Karlsruhe, as well as "PUSH!" A network of partners for university start-ups in the region of Stuttgart.

[13] Many regions participated in these two competitions; some of the concepts were even realised without (national) funding.

the principal network partners due to their ability to close the gap in the supply of risk-carrying capital, which distinctly profit-oriented venture capitalists (venture capital companies) do not provide for less growth-promising young and small (technology) companies.

However, despite the involvement of their network partners, active management support by regional venture capital companies is limited. They can only *offer* consultation, but have no means of direct influence of the portfolio company's management, in contrast to venture capital companies which participate directly in companies; moreover, they do not have the qualitative capacity of support concerning strategic issues such as the establishment and development of a (world-wide) sales organisation, the initiation of strategic alliances with important sales partners, or the acquisition of venture capital up to several millions, initial public offering, and so on.

In fact, the quality of support for portfolio companies, which is nevertheless possible, strongly depends on the quality of innovation network partners and their readiness for co-operation. However, due to the fact that network co-operation takes place on an informal level, it is difficult to evaluate the chances for the realisation of innovation networks in Germany which are aimed at the *consultation of regional venture capital companies' investees.*

11.5   Conclusions from German Experience which Is Relevant for Transition Countries

Although the establishment of innovation networks in Central and Eastern European transition countries is one of the main objectives of their respective innovation and technology policies, concrete measures for the support of network activities are characterised by the traditional political and legal framework conditions which still exist.[14] Three partial networks are important for (regional) venture capital companies:

- network relationships forming an interface between financing and scientific partners,

- network relationships between capital providers for innovation financing, and

- network links to public and private consultation agencies, promotional institutions, etc.

---

14 Among others, refer to: Walter/Bross (1997: 267ff.), and the illustration of the situation in Central and Eastern Europe shown there.

Regarding the establishment and development of regional innovation networks, regional venture capital companies can fulfil different functions, from active efforts to extend and deepen the network to the passive behaviour of a sporadic user of established network relationships.

Especially in technology, the establishment of an (informal) network of relationships with other venture capitalists and investors, research and consultation institutions, industrial companies, etc., is most important for venture capital companies, whether they are oriented towards trade and industry or towards promotion. This importance is not only due to easy access to interesting investment objects and the distribution of basic information for the selection of portfolio companies, but also to the possibility of efficient and appropriate support of the portfolio company's business activity. Co-operation with external partners can save considerable costs for every venture capital company, since their own personnel capacities are relieved. The establishment of a network aimed at the support of their investee companies would be of importance especially for promotion-oriented and less profit-oriented venture capital companies, all the more so since they should have easier access to other publicly run infra-structural institutions for entrepreneurial and innovative support than profit-oriented venture capital companies.

On the one hand, a regional venture capital company must make use of all options in order to minimise expenditure for the selection of appropriate portfolio companies; on the other hand, sufficient demand for their venture capital supply must be ensured. For promotion-oriented venture capital companies this should imply that the credit sector not only fulfils a pre-selection function but also contributes to the opening up of markets by arranging contacts between appropriate companies and venture capital companies. Founders' and technology centres, promotional investors, or consultation agencies are other institutions which are able to fulfil this function. Moreover, German experience shows that tax consultants and qualified auditors should be thoroughly informed about the financing offer of these venture capital companies.

Regional venture capital companies with these criteria are able to take charge of the equity financing of less growth-promising new foundations and of re-structured SMEs. Although each one of these corporate entities only has limited significance from a global economic or regional perspective, they nevertheless represent a considerable share. Regional venture capital companies also fulfil important functions in the cases of risk-carrying new foundations and re-structured SMEs:

- during the foundation stage or the first phase of re-structuring they ensure start-up financing until the company has reached a state of development which makes them attractive for private investors, and/or

- they play the role of co-investors; in this way the number of capitalists is increased, and the risk of the individual financier is reduced.

However, regional promotion-oriented venture capital companies only have limited possibilities to provide efficient management support, which is particularly important for companies with significant growth potential. Regarding this small group of young and small (technology) companies, and from a global economic point of view, close co-operation with venture capital companies would make sense; due to their competence and network of contacts, these could provide the "value-added" which would help the growth potential to become a real success on the market.

11.6   References

BROSS, U./WALTER, G.H. (1998): *Entwicklungsperspektiven des tschechischen Venture Capital Marktes. Bestandsaufnahme und Ansatzpunkte politischer Massnahmen.* Stuttgart: IRB-Verlag.

BUNDESVERBAND DEUTSCHER KAPITALBETEILIGUNGSGESELLSCHAFTEN - GERMAN VENTURE CAPITAL ASSOCIATION E.V. (BVK): *Several yearbooks.* Berlin.

EUROPEAN VENTURE CAPITAL ASSOCIATION (EVCA) (1999): *1999 Yearbook.* Zaventem: EVCA.

KULICKE, M. (1998): Die Finanzierung des Unternehmens, BIOTECHNOLOGIE-AGENTUR BADEN-WÜRTTEMBERG, LANDESGEWERBEAMT BADEN-WÜRTTEMBERG (Eds.): *BEN Biotechnologie-Existenzgründungs-Navigator. Ein Leitfaden für Existenzgründungen im Bereich der Biotechnologie.* Karlsruhe, Stuttgart: Landesgewerbeamt.

KULICKE, M./WUPPERFELD, U. (1996): *Beteiligungskapital für junge Technologieunternehmen. Ergebnisse eines Modellversuchs.* Heidelberg: Physica-Verlag.

LEOPOLD, G./FROMMANN, H. (1998): *Eigenkapital für den Mittelstand. Venture Capital im In- und Ausland.* München: Beck Juristischer Verlag

LESSAT, V./HEMER, J./ECKERLE, T./KULICKE, M./LICHT, G./NERLINGER, E. ET AL. (1999): *Beteiligungskapital und technologieorientierte Unternehmensgründung. Markt – Finanzierung – Rahmenbedingungen.* Wiesbaden: Gabler-Verlag.

LOMPE, K./KEHLBECK, H./SCHIRMACHER, A./WARNECKE, D. (1998): *Existenzgründungen, Risikokapital und Region.* Hannover: Nomos.

WALTER, G.H./BROSS, U. (1997): Transformation deutscher Erfahrungen beim Aufbau von Innovationsnetzwerken in Mittel- und Osteuropa, KOSCHATZKY, K. (Eds.): *Technologieunternehmen im Innovationsprozess. Management, Finanzierung und regionale Netzwerke.* Heidelberg: Physica-Verlag, pp. 267-291.

WUPPERFELD, U. (1996): *Management und Rahmenbedingungen von Beteiligungs-gesellschaften auf dem deutschen Seed-Capital-Markt.* Frankfurt am Main: Peter Lang.

Section IV: Innovation Networks and Regional Innovation Policy

12. Innovation, Interaction and Regional Development: Structural Characteristics of Regional Innovation Strategies

032 038

R11

R58

Andrea Zenker

(Germany)

12.1 Introduction

In the light of globalisation, increasing industrial specialisation, and growing competitive pressure, innovative developments and processes have reached a level of primary importance. New products and processes, as well as organisational and societal innovations, contribute to firms', regions', and nations' competitiveness. Considering innovative processes on a regional level provides insights into space-specific economic, historical, socio-political and cultural factors which form the background of industrial innovation processes. The region in which innovating firms are located represents their "home base" and here interactive relationships lead to the development of specific innovation patterns.

Regional innovation policy aims at the activation of innovation potential which has been unused so far, with the objective of improving competitiveness and success and, in so doing, to contribute to prosperity and growth. The main focus in this context is firms' innovation activity and their need for supportive services, contacts and interactions, as well as the establishment of an innovation-supporting environment. Therefore, initiatives for regional innovation promotion refer to theoretical considerations dealing with the concepts of networks, innovative milieus and learning regions. The points of departure, objectives, and the realisation of these initiatives are of different types and concern, for example, different fields of technology, the support of individual networks, or the arrangement of contacts and the development of co-operation between different regional actors.[1] The concept of regional innovation systems gives an analytical basis and a fundamental understanding of innovation processes on a regional level and can also be used as a reference framework for the evaluation of the present situation and for the orientation of

---

[1] In this contribution, regional actors include: firms, research and education organisations, providers of innovation-supportive services, as well as the people and organisations involved in innovation initiative processes, specific working groups, consultants, public agencies, etc.

innovation strategy. This paper includes short descriptions of conceptional elements which are illustrated with examples from practice.

12.2 Regional Innovation Systems

12.2.1 Conceptional Basis

The concept of innovation systems is based on the idea of the interactive innovation model according to the evolutionary approach. Whereas linear innovation models are based on the succession of basic and applied research, development, innovation, production and diffusion, the interactive innovation model implies that every phase of the innovation process contains feed-back loops with the other phases, and links to research and development as well as to the firm's knowledge base.[2] Schumpeter (1935) considered the process of innovations or of "creative destruction" as a fundamental impulse for the further development of capitalist systems. In the further development of this approach, representatives of the neo-Schumpeterian or evolutionary approach considered innovation as process-oriented development for which knowledge constitutes a fundamental factor (cf. Nelson/Winter 1974 and Kline/Rosenberg 1986). An interactive innovation model implies the continuous acquisition and appropriation of new knowledge, its combination with available knowledge and its application for specific problem areas within the firm.

The systemic character of innovations is associated with the idea that the innovation process is a result of social interaction between different actors. These characteristics, – innovation as an interactive and systemic process – represent the fundamental concept of innovation systems. According to Lundvall, a national innovation system comprises elements and relationships interacting in the production, diffusion and use of knowledge within a nation state. (Lundvall 1992: 2). Consequently, an innovation system is a social system which is permanently in contact with its environment; this continuous exchange is crucial for the introduction of new knowledge and new technology (Cooke 1998: 11). Besides private firms, innovation systems are constituted by research and educational institutions, providers of innovation support services, public administration, institutions for technology transfer and economic promotion agencies, as well as political institutions. According to the diversity of these elements, their relationships and their connections with their respective socio-cultural environment, there is no single "model" of a regional innovation system (cf. the contribution of Landabaso *et al.* in this volume).

---

2 *Cf.* the "chain-linked model of innovation" in Kline/Rosenberg (1986: 290).

12.2.2   Regional Innovation Networks

Innovation systems are being increasingly discussed not only for the national but also for the regional level.[3] The significance of a region lies primarily in its character as a "home base" for innovating firms, which offers an environment equipped with specific location factors. According to theoretical approaches referring to networks and innovative environments, common backgrounds facilitate the establishment of confidence as well as information and knowledge exchange and the readiness for co-operation and networks. Also from a political point of view, the regional level is gaining significance due to the fact that the strongest relationships to industry and other actors are found here (Landabaso/Youds 1999: 5).

Successful networks make use of the knowledge of all its members, and spatial proximity plays a particularly important role regarding the exchange of tacit knowledge by reason of easy face-to-face contacts and informal exchange. Consequently, the significance of physical proximity depends on the complexity and uncertainty of transactions, as well as on the amount of tacit knowledge to be exchanged (Senker 1995: 107 and Morgan/Nauwelaers 1999b: 7/8).[4] A common cultural and social background can lead to synergy effects in the innovation process through similar attitudes and opinions; it forms the framework conditions for a region-specific knowledge base which enables "interactive learning" (Asheim/Isaksen 1999: 3/4 and 7ff.). However, trust and openness are required from actors who are prepared to become part of a network. They must be sure that the knowledge introduced into the network will be treated with confidence, and that this will be reciprocal.[5]

The necessary pre-conditions therefore are specific norms and rules, reciprocity, and trust, in short: social capital. Consequently, firms' knowledge base and their learning capacities are not only the pre-condition for the absorption and processing of external knowledge, but they also serve as an "admission ticket" to the network; at the same time, they contribute to the competitive advantage of a firm (Rosenberg 1990 and Hicks 1995: 409ff. treat this questions from the viewpoint of firms' R&D activities).

---

3   Regarding the concept and elements of regional innovation systems, refer to Braczyk *et al.* (1998), Muller *et al.* (1994), and Koschatzky (1997), among others.

4   Morgan/Nauwelaers (1999b: 8) note that the meaning of the factor proximity has changed. Whereas it was formerly seen as a cost-reducing factor, it is presently discussed as a support of learning processes. According to this, spatial proximity enables the exchange of tacit knowledge, whereas the knowledge exchanged in national and global networks has a more explicit tendency: "Networks can also be international or even global, and the frequency of these wider networks is growing rapidly, but normally wider networks will tend to be more formalised and oriented toward the exchange of codified knowledge." (Lundvall/Borrás 1997: 105).

5   As a result, firms are always in a conflictual situation between co-operation and competition: "Collaboration, for firms, means a trade-off between access to greater resources against the potential for loss of valuable proprietary information to competitors." (Cooke 1998: 5).

12.2.3 Knowledge and Learning on a Regional Level

According to interactive approaches, knowledge and its extension through learning are fundamental factors in modern economies: "...the most fundamental resource in the modern economy is knowledge and, accordingly, [.] the most important process is learning." (Lundvall 1992: 1). Firms learn from their own experience, as well as from and with other actors with whom they co-operate and exchange knowledge (Cooke 1998: 8). Knowledge and learning are transferred from the level of an individual firm to that of the whole network and thus fulfil the function of a fundamental and strategic factor for the whole regional system. Consequently, the competitiveness of "learning regions" depends less on the availability of physical resources, but on specific regional competence,6 which, based on their regional context and specialisation, is reached by the acquisition and application of knowledge, as well as by learning processes: "The relative performance of regions as well as the relative performance of firms is merely the superficial expression of a deeper competition over competences" (Lawson 1997: 15). Due to the fact that regional competence is not only established and deepened through the internal acquisition of knowledge but also through inter-regional interaction, the openness of a region is an important pre-condition for its competitiveness and its success. The engagement in supra-regional networks allow the "home environment" to benefit from the influx of new knowledge into the regional network (Cooke 1998: 10). However, this also implies that a region, in the course of its development process towards a "learning region", should react to external changes in a flexible way, and that it also should be able to "forget", i.e. that structures and methods which were adapted to formerly valid conditions should be replaced by new ones in the case of changing framework conditions.7 Instead of networks *per se*, dynamic networks encourage learning and innovation (see also Morgan/Nauwelaers 1999b: 11); otherwise, inflexible structures and effects of persistence could ensue.

Consequently, in comparison with network and milieu approaches, the significance of interaction is more related to knowledge and competence. In contrast to physical goods, the value of competences of a "learning region" are not diminished through use; in fact, they are improved and extended through intensive joint application and utilisation with partners (Lawson 1997: 15), so that actors in "learning regions" can extend their own knowledge base, as well as that of their region, through co-operation. In learning regions, innovation occurs in a continuous way rather than

---

6 Lawson (1997: 10), when discussing the transfer of the competence aspect from the firm to the regional level, concludes: "...although firms and regions are not the same things, both are ensembles of competences which emerge from social interaction and so there appears to be no reason at all why the competence perspective should not be as equally relevant to the study of the region as to the study of the firm."

7 See Hassink (1997: 280). Referring to Stiglitz, Cooke (1998: 13) talks about "learning by learning" in this context, which includes permanent self-monitoring and adaptation to modified framework conditions.

through individual points, and is then integrated into the production process; moreover, firms are becoming increasingly specialised (Smith 1995: 80) in core competence. Consequently, firms rely more on external knowledge in order to complete their own competences.

The concepts of the regional innovation system and of the "learning region" are linked by the strong emphasis on knowledge, learning, and competence. On the one hand, a learning process is continuously happening in a regional innovation system due to the necessity of the constant acquisition, diffusion, and application of knowledge for the purpose of realising innovations. On the other hand, the innovation system presupposes the establishment of a regional knowledge base and the utilisation of existing potentials (Landabaso/Youds 1999: 2/3, and Braczyk/Heidenreich 1998: 423). Emphasising the character of knowledge and learning in a specific regional context, regional innovation policy is brought into play at this point, with the concept of regional innovation systems as a reference framework for the different options of political design (see also Asheim/Isaksen 1999: 14/15).

12.3   Regional Innovation Initiatives – Concepts, Strategies, and Elements of Success

In the following, the conceptional perspectives cited above will be illustrated through examples of regional innovation initiatives. For this, partial results of the project "regional distribution of innovation and technology potentials in Germany and in Europe"[8] will be used. This project is aimed at the analysis of innovation at regional scale, the deduction of reasons for regional innovation disparities, and, finally, at drawing conclusions for innovation policy. These questions were approached from two sides by the project team: firstly, by starting on a "macro level" by means of an inventory of available indicators of regional innovation, secondly, on a "micro level" by the attempt to gain additional information about the ongoing processes in the individual regions through insights into regional innovation initiatives. Different innovation initiatives from selected German and European regions were studied through an analysis of the available material. The results are based on an analysis of documents about the RTP, RIS, and RITTS[9] measures, supported by

---

8   This project is presently being carried out for the Federal Ministry of Education and Research (cf. BMBF 2000: 86-97) by the Fraunhofer Institute for Systems and Innovation Research (FhG-ISI, Karlsruhe), by the Lower Saxonian Institute for Economic Research (NIW, Hanover), by the German Institute for Economic Research (DIW, Berlin), and by the Kiel Institute of World Economics (IfW, Kiel).

9   RTP: Regional Technology Plan. RIS: Regional Innovation Strategy. RITTS: Regional Innovation and Technology Transfer Strategies and Infrastructures. The RTP programme, which started in 1994, encouraged the development of a regional technology plan aimed at the strengthening of regional innovation potential through improved research, technology development and innova-

the European Commission, Directorates-General Regional Policy and Enterprise, as well as on the analysis of further initiatives in German regions.[10]

12.3.1 Concepts of Regional Innovation Initiatives

Similar patterns are found in the methodology of all innovation initiatives which were studied: indicators were collected regarding the regional context, firms' needs for innovation support, as well as regarding the supply of such services by institutions located in the respective regions in order to analyse the present situation in a region, to define a development path, and to derive an innovation strategy on this basis. This information is gathered through quantitative and/or qualitative analyses and conclusions. Data about the regional location conditions, regional infrastructure, and structural features of regional firms are collected in order to analyse the regional economic structure. Included are, for example, the regional structure of manufacturing and service firms, of research and education institutions, general structural data regarding the firms and plants (size, structure of employees, age, sectors), as well as their characteristics regarding technology, research, and development. After comparing the information resulting from this investigation, the regional production and innovation system is characterised, and first indications concerning strengths and weaknesses can be derived. On this basis, action plans are worked out, and the whole process is presented to the public through workshops, panel discussions, etc. Target groups are defined and sensitised for the identified topics. Finally, pilot projects are formulated and carried out. In order to control the progress of the study and to derive "good practices", the European Union intends to assist the process through monitoring and evaluation. The exchange of experiences with other initiatives, as well as benchmarking, help to further optimise the process of an innovation initiative.

---

tion in a regional context. In 1996, the RTP measures were transformed into RIS, regional innovation strategies. The objective of RITTS is the support of technological capacities in SMEs through technology transfer. Whereas the RIS programme is funded by Directorate-General (DG) Regional Policy (former DG XVI regional policy and cohesion) due to article 10 of the European Fund for Regional Development (EFRE), the RITTS programme belongs to the sphere of responsibility of Directorate-General Enterprise (former DG XIII Telecommunications, information market, utilisation of research results) and is financed by the innovation programme. Both programmes are partly funded by the European Community, further funding must be applied for by the regions which aspire to participation. The objectives of the programmes RIS and RITTS are "• to improve the capacity of regional actors to develop policies which take into account the real needs of the business sector and the strengths and capabilities of the regional innovation system; • to provide a framework within which both the European Union, the Member States and the regions can optimise policy decisions regarding future investments in RTD, innovation and technology transfer initiatives at regional level." (European Commission 1997: 8). Regarding characteristics, differences and common features of these two programmes, refer to European Commission 1997: 7ff.

[10] The project team was supported by Dr. Busch, DG Enterprise (Luxembourg), and Dr. Landabaso, DG Regional Policy (Bruxelles).

Although the realisation of individual steps can vary between the respective initiatives, all of them are oriented towards the concept of regional innovation systems and follow a bottom-up approach including interaction, placing firms and their innovation activities at the centre of attention.[11] If the analysis of the present situation of innovation support shows compatibility between the supply and demand of these services, and therefore an efficient partial system, then almost no (supportive) interference is necessary. However, frequently so-called "fragmented innovation systems" can be identified: here, although individual elements largely exist, their co-operation with each other is not sufficient (cf. the contribution of Landabaso *et al.* in this volume). In this case, innovation strategies include the organisation of interaction between the individual elements. If innovation systems prove not only to be fragmented but, in addition, "incompatible" (i.e. no development of interactive relationships is possible due to a lack of co-operation partners), then the point of departure of political measures lies in the elements themselves and their interaction. The appropriate innovation strategies are aimed at converting the fragmented regional systems into functioning, dynamic, and comprehensive systems where the basic requirements for successful innovation projects are present.

The final stage of regional innovation initiatives is dedicated to the implementation of measures. To begin with, pilot schemes are worked out and realised; in a second step, experiences and results are integrated into further project plans. Experience shows that the implementation process should start as quickly as possible so that the relevance and success of the strategy development process can be proven to participants (Boekholt *et al.* 1998: 4 and 82). Otherwise, disappointment about the absence of success following their commitment would lead to a lack of motivation to further participate in the present or in other initiatives.

12.3.2   Strategies of Regional Innovation Initiatives

The strategies derived from the analysis of the respective regional situations have specific starting-points, which refer to the individual actors, their interactions, and an innovation-supportive environment. One of the first tasks of a regional innovation initiative is the definition of an organisational and management structure,[12] and the definition of the region in which the initiative is performed. Due to the fact that different spatial dimensions can be affected according to the point of approach, the

---

[11] Nauwelaers/Morgan (1999: 231) formulate the role of firms in a political approach which interactively connects supply and demand: "...it is absolutely crucial in this "interactive" policy perspective that the enterprises themselves realise that they are at the centre of the problem, and also at the core of the solution. When the enterprises have progressed along this path, it becomes possible for them to be associated with the exercise and to play an active role in it."

[12] In general, the management of innovation initiatives is carried out by public or semi-public institutions such as chambers of commerce and industry, technology transfer agencies, regional administration or innovation agencies.

region in which the initiative is performed varies according to the individual objective. The establishment of consensus among the regional actors is crucial from the beginning of the process, to help them to identify themselves to a far-reaching degree with the measures to be taken, in order to motivate them to participate, and to establish a good foundation for co-operation. The involvement of firms is of particular interest in order to raise the degree of integration and acceptance of the programme, and in order to align the developed strategies with precise industrial requirements.

The study of different innovation initiatives leads to an exemplary identification of the following *strategic elements*, which – according to defined objectives – are embedded in a global strategy and realised in pilot schemes or projects.

Establishment of awareness and regional identity
This refers to the establishment of awareness regarding the importance of innovation for the competitiveness of firms and the region. In order to create awareness for innovation in regional firms, the regional innovation strategy of Aragón, for example, proposed public discussions about innovation and competitiveness, information on innovation and technology-relevant questions, consultation concerning innovation programmes, assistance with the handling of applications, and, last but not least, a directory of regional innovation-supportive services. The RITTS Nord-Pas de Calais conceived "Clubs-Innovation" for managers and middle managers, and "Campus-Innovation" targeted at project promoters. Some of the activities entitled "innovation competence" in the framework of RITTS Overijsel was the broadcasting of an innovation campaign by regional television, whereas "The Welsh Innovation Challenge Competition", was described as a "flagship project" in the action plan of the RTP Wales. In view of a tighter innovation-bound concerted action, a forum for regional innovation emerged from the RITTS West Norway, which is responsible for the co-ordination of regional innovation activities, as well as for the monitoring and evaluation of the priority projects included in the action plan (Europäische Union, no year given: 34/35; Agence Régionale de Développement Nord-Pas de Calais 1999: 4; Province of Overijsel, no year given: 9; Welsh Development Agency, no year given: 10/11; No author given, no year given: RITTS Western Norway: 8/9).

The creation of regional consciousness represents a strategic element of the RIS Western Macedonia. During so-called participative workshops on different subjects (with an expert and interested members of the advisory board), attention was drawn to the importance of co-operation on a regional level. The importance of regional innovation policy was emphasised in one of the strategic elements of the RTP Lorraine; here, a "regional charter for innovation and technology transfer" was adopted. The RITTS Aachen also included, as part of their strategy, the planning of a long-term innovation and technology transfer, as well as consensus building for this strategic planning. The region of Weser-Ems aimed at creating a common regional awareness with its slogan "We are growing – together" (Europäische Union, no year given: 18/19, and 50/51, No author given, no year given: RITTS Aachen: 4, and RIS Weser-Ems Interest Group 1998).

Support of individual actors and their interactions
Strategies in this area are mostly focused on the providers of innovation support services and/or on the demand side, partly also on the administrative or political area. In this respect, the creation and improvement of exchanges between the individual elements of the innovation system is the main objective. Many of these (sub-) strategies include programmes for the establishment of co-operation between industry and research, or between industrial and educational institutions. Further strategic elements are the promotion of existing technological capacities in the region, for example by working out specific promotional programmes, or the organising of access to knowledge outside the region. Other initiatives worth mentioning are the creation of a financing basis for innovation projects or specific measures for special industrial groups such as the promotion of young enterprises or start-ups, the introduction of new technology into specific industrial sectors, or the promotion of the internationalisation of economic activities. Last but not least, the support of marketing success reached through the development of innovation is one of the listed strategic elements.

In this area, a great number of realised examples can be cited from the studied initiatives. The strategies of some of the initiatives include networking between industry and educational institutions, others develop programmes for the support of start-ups (e.g. RTP Lorraine), or support specific key technology fields. The Regional Technology Plan Halle-Leipzig-Dessau is an example for the support of regional firms' supra-regional orientation; the framework of this programme encourages interregional co-operation between the "Länder" Saxony and Saxony-Anhalt, based on networks between firms, research institutions, and public administration. A comparison between endogenous strengths and weaknesses and supra-regional trends was part of the RIS programme for the Weser-Ems region, whereas the region of Limburg (NL) emphasised knowledge and globalisation in its RTP programme (Europäische Union, no year given: 18/19 and 20/21; RKW Sachsen/Institut für Strukturpolitik und Wirtschaftsförderung Halle-Leipzig e.V., no year given: 1; RIS Weser-Ems Interest Group 1998: 11ff.; Provincie Limburg 1996: 21ff.).

Information and communication
Innovation activities are furthermore supported by the bundling and distribution of information within and outside the region and by the support of communication within the regional system. This includes the supply of information for actors, the organisation of information events and workshops, etc., on the one hand, and, on the other hand, the development of a regional marketing concept which is suitable for actors on both an internal and external level. Presentations of the initiatives at na-

tional and international trade fairs and similar events improve the visibility of the region.

To promote the "industrial culture" and access to enterprise networks, the action plan of the RITTS programme for Milano comprises, among others, a project for the establishment of an intranet between groups of SMEs, whereas the region of Overijsel has designated its "ICT Programme" for a better utilisation of information and communication technology. Improved communication between government and citizens is planned, as well as the publication of calls for tenders on the internet. An example of regional marketing and the supply of information about specific groups of firms is provided by the "TechnologieRegion Karlsruhe" and the cyber forum of the TechnologieRegion. The TechnologieRegion is a voluntary alliance of different urban and rural districts, aimed at common internal and external marketing of regional firms. The cyber forum describes itself as a "virtual industrial park" which supports firm foundations from the multimedia sector by providing information (Provincia di Milano 1999: 13; Province of Overijsel, no year given: 14; TechnologieRegion Karlsruhe, no year given; Cyberforum).

12.3.3 Success Factors of Regional Innovation Initiatives

General statements to measure the success of regional innovation initiatives are subject to many difficulties and uncertainties. This is not only due to the great number and diversity of the specific initiatives but also their medium to long term character, which hinders the short-term visibility of success. Moreover, the relation between a political measure and output factors is not linear and univariate, but of a most complex character. In the short and medium term, the success of an initiative can only be measured by the achievement of self-imposed goals, whereas changed macro- indicators such as the number of foundations or of patent applications, or the increase in employment figures of a region are not assessable in the short term.[13]

Consequently, regional innovation initiatives should be considered as a process oriented towards a sustainable effect in the region. This should be considered by the management team in order to avoid disappointment if immediate effects fail to appear. Another important element is the compatibility of developed strategies with regional framework conditions. The question is to what degree strengths and weak-

---

[13] Morgan/Nauwelaers (1999a: XVI/XVII) emphasise this: "...regional innovation policies do not satisfy the criterion of conventional regional policy, namely short-term job creation", and stress the indirect effect of innovation policy on the regional employment situation: "...innovation policies do target the problem of employment because they tackle the question of the competencies of firms and regions. And there is no way around the fact that the creation of durable jobs depends on a strong economy. [...] Regional innovation policies are not a *direct* solution to the job creation imperative, and should be complemented by other public policies, notably targeting local employment possibilities..." (Nauwelaers/Morgan 1999: 237)

nesses were taken into account for the development of the strategy, and whether regional consensus could be pooled for the resulting strategy. A key function in the whole process is held by the management structure, as well as by the moderators, whose task is the motivation of regional actors to participate in the initiative (European Commission 1997: 9, and 19ff.). Their main task is to build regional consensus concerning the importance of innovation, innovation policy and the initiative, to spread the initiative's objectives and advantages in the region, and thus to win regional actors for the initiative and encourage them to identify themselves with the plan (Landabaso/Youds 1999: 9). This can be realised, for example, by information disseminating events, by presentations at the beginning of and during the process, by workshops with specific target groups, by information brochures, or by integrating mass media (European Commission 1997: 30). The development of confidence and of social capital in the region[14] are very important factors, which form the foundation for co-operation and networks. Last but not least, openness towards the experience of other regions, as well as the participation in platforms for exchange and networks facilitate regional learning and help with the transfer of experience from other regional initiatives. Another factor which promises success is the demand orientation of the initiative.[15] According to experience in innovation initiatives realised up to the present, firms are often not aware of their concrete need for assistance and therefore have difficulties with the formulation of their needs for support. Awareness concerning the issue of innovations, as well as to the possibilities of co-operation on a regional level, can help firms with the definition and formulation of their objectives.

In contrast to this, different factors can endanger regional innovation initiatives. These are for example: insufficient focussing on firms, as well as confrontations about power within a region, or the preservation of inadequate structures. Finally, the inefficient use of funds, i.e. granting investments in an area outside of the centre of the initiative, or too long a stretch of time between strategy development and the implementation of first measures, can lead to de-motivation of the initiative's participants. The success of an innovation initiative can be measured in the medium term by ascertaining whether innovation planning has gained significance on firm and regional level, or whether the influence of policy making could be augmented (Landabaso/Reid 1999: 35ff. and Boekholt et al. 1998: 35ff.). However, in this area, measurements according to "fixed", uniform indicators must give way to a "with-and-without" comparison. Given a lower starting position at the beginning of the initiative, the change of attitudes in a region can lead to tremendous learning ef-

---

14 Morgan/Nauwelaers (1999b: 4) distinctly emphasise the importance of this element: "Traditional definitions of innovation make too much of the tangible technical dimensions and too little of the intangible social, organisational and relational dimensions."

15 Whereas, in many regions, innovation policy has stressed the support of infrastructure up to now, recent experience has shown that in particular an exchange between suppliers and demanders of innovation support is an important pre-requisite for the compatibility of supply and demand.

fects, even if the objectively reached target remains below that of other regions which had a higher starting position (Boekholt *et al.* 1998: 19).

The description of possible (short and medium-term) success factors of regional innovation initiatives reveals difficulties in operationalisation and measurement of success. It is difficult and in some points even impossible to record the listed success factors. Nevertheless, indirect indicators can be defined which allow statements about a regional innovation initiative to be made. These include, for example, the degree of integration of regional firms, the demand orientation of service support, the targeting of action plans at the business sector, and the co-operation relationships which have come into existence and which will have an indirect influence on economic success through learning effects and innovations. Finally, the continuation of the initiative beyond the formulation of a strategy, i.e. the definition and implementation of pilot schemes, can be listed as a success factor (cf. Boekholt *et al.* 1998: 50ff.).

12.4   Conclusion

Local innovation conditions and innovation patterns are determined by the general socio-cultural and economic framework conditions, the individual actors and their interactions, as well as the knowledge base including codified and tacit knowledge. Consequently, no regional innovation system resembles another; instead, each one has specific characteristics, and thus follows its own development path. The same is true for the analysed regional innovation initiatives, which are just as heterogeneous. Despite a high degree of diversity, certain common aspects can be found: implicitly or explicitly, all of them are based on the theoretical concept of interactive and systemic innovation processes, on the importance of co-operation and network relationships, and the relevance of the regional environment. The significance of knowledge and learning is emphasised, as well as the embeddedness of initiatives into the regional context. Important elements are the regional actors, network moderators, manpower and the population, as well as consensus between them, which is necessary for a joint discussion process to lead to the realisation of an initiative. From a political perspective, these characteristics are being discussed within the reference framework of regional innovation systems.

What are the consequences of these aspects for innovation policy? As seen from the examples, innovation support has developed from the (supply-oriented) support of infrastructure to the (interaction-oriented) support of innovation, which gives special attention to the demand for innovation-supporting services. In this way, innovation support follows the latest theoretical findings which emphasise the importance of interaction and network relationships. Moreover, to support innovations means quite more than technology support since it includes the whole innovation process. Consequently, regional innovation policy aims at the stimulation of industrial innovation, as well as at the establishment of appropriate basic requirements, and at enabling the regions to qualify for the successful realisation of innovation projects. As was shown by the exemplary regional innovation initiatives, the possibilities for policy action are manifold but also very complex, since prominence is given not only to the support of individual actors but also to their interactions, as well as to the so-called "soft factors" which include social capital, regional knowledge, and learning. In the light of the specific regional framework conditions and knowledge base, each region has to find its own way, since: "...there is no such a thing as an optimal system of innovation: those systems are place- and time-specific, and in constant evolution. Therefore, innovation policies can only be built and evaluated in relation to the situation to which they refer: there is no one-size-fits-all model for policy-building in this area of innovation support." (Nauwelaers/Morgan 1999: 236). On the one hand, this represents a specific challenge for innovation policy; on the other hand, it opens up specific chances for innovation-oriented regional development.

12.5   References

AGENCE RÉGIONALE DE DÉVELOPPEMENT NORD-PAS DE CALAIS (ARDNPC) (1999): *Synthèse du rapport conclusif.* http://www.ardnpc.org/europe/ritts-/Synthese120199.htm (March 2000).

ASHEIM, B./ISAKSEN, A. (1999): *Regional Innovation Systems: The Integration of Local 'Sticky' and Global 'Ubiquitous' Knowledge.* Paper presented at the NECSTS-99 Conference on 'Regional Innovation Systems in Europe'. San Sebastian, 30 September – 2 October.

BMBF [BUNDESMINISTERIUM FÜR BILDUNG UND FORSCHUNG] (Ed.) (2000): *Zur technologischen Leistungsfähigkeit Deutschlands. Zusammenfassender Endbericht 1999.* Bonn: BMBF.

BOEKHOLT, P./ARNOLD, E./TSIPOURI, L. (1998): *The Evaluation of the Pre-Pilot Actions under Article 10: Innovative Measures Regarding Regional Technology Plans.* http://www.innovating-regions.org/library/library_3_1.html (October 1999).

BRACZYK, H.-J./COOKE, P./HEIDENREICH, M. (Eds.): *Regional Innovation Systems.* London: UCL Press.

BRACZYK, H.-J./HEIDENREICH, M. (1998): Regional governance structures in a globalized world, BRACZYK, H.-J./COOKE, P./HEIDENREICH, M. (Eds.): *Regional Innovation Systems.* London: UCL Press, pp. 414-440.

COOKE, P. (1998): Introduction: origins of the concept, BRACZYK, H.-J./COOKE, P./HEIDENREICH, M. (Eds.): *Regional Innovation Systems.* London: UCL Press, pp. 2-25.

CYBERFORUM. http://www.cyberforum.de/ (October 1999).

EUROPÄISCHE UNION, GENERALDIREKTION Regionalpolitik und Kohäsion (no year given): *Innovative Massnahmen unter Artikel 10 (EFRE). Förderung der Innovation*. Brüssel.

EUROPEAN COMMISSION, DG XVI & DG XIII (1997): *RIS/RITTS Guide*. http://www.innovating-regions.org/library/library_4_2.html (October 1999).

HASSINK, R. (1997): Localized Industrial Learning and Innovation Policies, *European Planning Studies*, 5, pp. 279-282.

HICKS, D. (1995): Published Papers, Tacit Competencies and Corporate Management of the Public/Private Character of Knowledge, *Industrial and Corporate Change*, 4, pp. 401-424.

KLINE, S.J./ROSENBERG, N. (1986): An Overview of Innovation, LANDAU, R./ROSENBERG, N. (Eds.): *The Positive Sum Strategy. Harnessing Technology for Economic Growth*. Washington: National Academy Press, pp. 275-305.

KOSCHATZKY, K. (1997): Innovative Regional Development Concepts and Technology-Based Firms, KOSCHATZKY, K. (Ed.): *Technology-Based Firms in the Innovation Process. Management, Financing and Regional Networks*. Heidelberg: Physica-Verlag, pp. 177-201.

LANDABASO, M./REID, A. (1999): Developing Regional Innovation Strategies: The European Commission as Animateur, MORGAN, K./NAUWELAERS, C. (Eds.): *Regional Innovation Strategies. The Challenge for Less-Favoured Regions*. Regions, Cities and Public Policy Series. London: The Stationary Office/Regional Studies Association, pp. 19-39.

LANDABASO, M./YOUDS, R. (1999): Regional Innovations Strategies (RIS): the development of a regional innovation capacity, *SIR-Mitteilungen und Berichte*, 27, pp. 1-14.

LAWSON, C. (1997): *Towards a Competence Theory of the Region*. ESRC Centre for Business Research, University of Cambridge, Working Paper No. 81. Cambridge.

LUNDVALL, B.-Å. (1992): Introduction. In: LUNDVALL, B.-Å. (Ed.): *National Systems of Innovation. Towards a Theory of Innovation and Interactive Learning*. London: Pinter Publishers.

LUNDVALL, B.-Å./BORRÁS, S. (1997): *The Globalising Learning Economy: Implications for Innovation Policy*. Reports based on contributions from seven projects under the TSER programme. Ed.: COMMISSION OF THE EUROPEAN UNION, DG XII/G. EUR 18307EN. Brussels.

MORGAN, K./NAUWELAERS, C. (1999a): Introduction, In: MORGAN, K./ NAUWELAERS, C. (Eds.): *Regional Innovation Strategies. The Challenge for Less-Favoured Regions*. Regions, Cities and Public Policy Series. London: The Stationary Office/Regional Studies Association, pp. XV-XVIII.

MORGAN, K./NAUWELAERS, C. (1999b): Regional Perspective on Innovation: From Theory to Strategy, MORGAN, K./NAUWELAERS, C. (Eds.): *Regional Innovation Strategies. The Challenge for Less-Favoured Regions*. Regions, Cities and Public Policy Series. London: The Stationary Office/Regional Studies Association, pp. 1-18.

MULLER, E./GUNDRUM, U./KOSCHATZKY, K. (1994): *Horizontal Review of Regional Innovation Capabilities*. Karlsruhe: ISI.

NAUWELAERS, C./MORGAN, K. (1999): The New Wave of Innovation-Oriented Regional Policies: Retrospect and Prospects, MORGAN, K./NAUWELAERS, C. (Eds.): *Regional Innovation Strategies. The Challenge for Less-Favoured Regions*. Regions, Cities and Public Policy Series. London: The Stationary Office/Regional Studies Association, pp. 224-238.

NELSON, R./WINTER, S.G. (1974): Neoclassical vs Evolutionary Theories of Economic Growth. Critique and Prospectus, *Economic Journal*, December, pp. 886-905.

NO AUTHOR GIVEN (no year given): *RITTS 038 - Der Abschlussbericht der regionalen Innovations- und Technologietransferstudie RITTS in der Region Aachen*. http://www.innovating-regions.org/library/library_2_1_de.html (October 1999).

NO AUTHOR GIVEN (no year given): RITTS *Innovation in Western Norway, Final Report*. http://www.innovating-regions.org/library/library_2_1_no.html (March 2000).

PROVINCE OF OVERIJSEL, DEPT. OF ECONOMICS, ENVIRONMENT AND TOURISM (no year given): *RIS+ Proposal Province of Overijsel NL*. http://www.innovating-regions.org/library/library_2_1_nl.html (March 2000).

PROVINCIA DE MILANO (1999): *Strategic Programme for the Development and the Support of Innovation in the Province of Milan 1999-2001*. http://www.innovating-regions.org/ActionMap/regionPage7.html (March 2000).

PROVINCIE LIMBURG (1996): *Regional Technology Plan Limburg*. Working document. Maastricht.

RIS WESER-EMS INTEREST GROUP (1998): *Regional Innovation Strategy Weser-Ems*. Oldenburg.

RKW SACHSEN/INSTITUT FÜR STRUKTURPOLITIK UND WIRTSCHAFTSFÖRDERUNG HALLE-LEIPZIG E.V. (no year given): *Regionale Innovationsstrategie Halle-Leipzig-Dessau*. Ein gemeinsames Projekt des Landes Sachsen-Anhalt und des Freistaates Sachsen mit Unterstützung der Europäischen Union. http://RIS.Halle-Leipzig-Dessau.org/home/ris.html (October 1999).

ROSENBERG, N. (1990): Why do firms do basic research (with their own money)?, *Research Policy*, 19, pp. 165-174.

SENKER, J. (1995): Networks and Tacit Knowledge in Innovation, *Economies et Sociétés*, Série Dynamique technologique et organisation, 2, pp. 99-118.

SMITH, K. (1995): Interactions in Knowledge Systems: Foundations, Policy Implications and Empirical Methods, *STI Review*, No. 16, pp. 69-102.

TECHNOLOGIEREGION KARLSRUHE (no year given): *10 Jahre TechnologieRegion Karlsruhe.*

WELSH DEVELOPMENT AGENCY (no year given): *Wales Regional Technology Plan. An Innovation and Technology Strategy for Wales.* Action Plan.

13. Innovation Networks and Industrial Research in the New Federal States: Perspectives of Economy and Structural Policy[*]

Herbert Berteit

13.1 Introduction

Today, the production factor "knowledge" has become the decisive driving force for economic structural change, with manifold ecological and social effects. The faster companies succeed in transforming new knowledge into novel or improved products or proceedings, and the faster they generate innovation, the more their competitiveness is strengthened in a sustained way. This is becoming more and more important due to the increased closeness of international markets and to shortened development cycles.

Consequently, a growing number of companies, research institutions, and a large variety of regional, national and international actors are trying to create innovation in their respective fields of interest, and in accordance with their specific objectives; they therefore bundle resources in co-operation relationships. Rapid technological progress and the growing complexity of customers' requirements lead to an increase in the intensity of competitive pressure.

This development becomes more and more significant in view of the bundling of regional competence on the one hand, and of the organisation of regions' competitiveness on the other hand.

In this process, the establishment of regional networks is particularly important. As was derived from the experience of empirical research on networking gained up to the present, the establishment of regional networks, which were initiated by indi-

---

[*] The paper summarises essential contents of an expertise entitled "Framework conditions for innovation networks in the new Federal States", which was carried out by SÖSTRA Berlin and SV-Wissenschaftsstatistik Essen, commissioned by the Federal Ministry of Economy and Technology; it was concluded in the middle of 1998.

224

vidual regional actors within developing regional economic systems, only advances slowly. In most of the cases, governmental support was granted to a greater or lesser degree, in different ways, and in different phases.

Whereas, in the former Federal States, networks between industry, scientific institutions and other regional actors, which were aimed at the achievement of innovative results, were able to develop over time, existing connections were broken off in East Germany due to the transformation from a directed economy to a free market economy, without the simultaneous establishment of new innovative networks. Industry, and with it industrial research, was in danger of collapse.

In the light of this, the Federal Ministries of Education and Research and of Economy and Technology, as well as the Federal States, committed considerable financial resources to the new Federal States, in order to promote industrial research and innovation networks. By the end of 1998, an amount of approximately 7.5 billion DM had been reached. In 1999, the Ministry of Education an Research launched the programme "InnoRegio" as an important initiative for the new Federal States. Furthermore, the Ministry of Economy has recently decided to support the establishment of innovative networks in the whole Federal Republic with an initiative called "Innonet".

These initiatives are directed at regional actors in the new Federal States and aim at supporting regional innovation potential, in view of a sustained improvement in terms of incremental value, competitiveness and employment on the side of the regions. The intention is to encourage both the formation of new networks and the expansion of recently implemented innovative networks, including the utilisation of existing potential.

This also includes the support of:

- the introduction of new quality standards and conditions of flexibility,
- the improvement of opportunities for the innovation-oriented modernisation of branches, or the structural change of these, and
- an increase of technical competence.

The long-term establishment of stable innovation networks between companies with different size structures and external research institutions closely related to the economy, as well as university and private institutes, offers good chances for opening up considerable potential and internal competitive reserves through a so-called "transfer via knowledge".

In the following, some aspects of the establishment of innovation networks in the new Federal States are shown, as well as the actual state of development of industrial research. Both fields are closely linked in view of their further development, especially in the new Federal States.

13.2 Innovation Networks

13.2.1 Concerning the Theoretical Background

The interplay of economy and science has advantages for both sides: on the one hand, science is given the opportunity to analyse technological problems in the light of the latest technology; for many scientists, such an activity is also considered as a challenge for another qualification (e.g. the obtaining of a doctorate).

The economy, on the other hand, has improved opportunities to utilise scientific findings and, at the same time, to make scientists aware of entrepreneurial requirements.

In this way, scientists acquire a new position as the "pulse generators" of application-oriented brainwork in their respective scientific institutions.

This is sufficiently proven by the experience gained through research on international co-operation, especially regarding successful "industrial districts" in industrialised Western countries, and the results of research on the "Mittelstand". Through co-operation, resources and potential complement each other, and the transfer of knowledge occurs. New conditions of flexibility and quality ensue. Due to the integration of knowledge from university and private research institutes, an "innovation-oriented modernisation" of the economy is supported, and a decisive expansion of technical competence takes place. "Quick and sustainable success requires carefully directed impulses and appropriate public support, which should not be restricted to helping overcome bottlenecks in the development of specific firms, but which should also deal with the problems of co-operation between companies in an intensified way" (Semlinger et al. 1993).

Particularly good results were and still are achieved wherever steady innovation networks have been developed over years in a goal-directed way, between companies of different size structures on the one hand, and public and external R&D institutions on the other hand.

Recent analyses dealing with growth theory show that the potential level of productivity also depends on the individual co-operation relationships which companies work in. Impulses for growth and employment are increased due to technological co-operation with other companies, relationships with customers and suppliers, interlocking capital arrangements, as well as co-operation with university and/or public research institutions (Hutschenreiter/Penender 1994).

Examples of this are the relationships with suppliers and with customers of the automobile industry, as well as relationships between manufacturers of capital goods and users.

In view of the dynamic of structural change, and in order to preserve and extend their competitiveness, especially small and medium-sized enterprises are interested in utilising possible synergies of innovation networks. These also outweigh risks such as the possible loss of technological or organisational independence.

Both modern economic theory and empirical research clearly show that economic growth and employment are not statically borne by traditional production factors, but that the efficient combination of these factors has a great influence on the competitiveness and innovation capacity not only of a company, but also on the whole economic and societal system (Messner/Meyer 1993).

By analogy with Gemünden and Heydebreck, innovation networks are represented above all as networks dominated by companies, which integrate technology partners, partners from different industrial sectors and from different markets, and regional partners (cf. Figure 13-1):

Figure 13-1: **Networks dominated by companies**

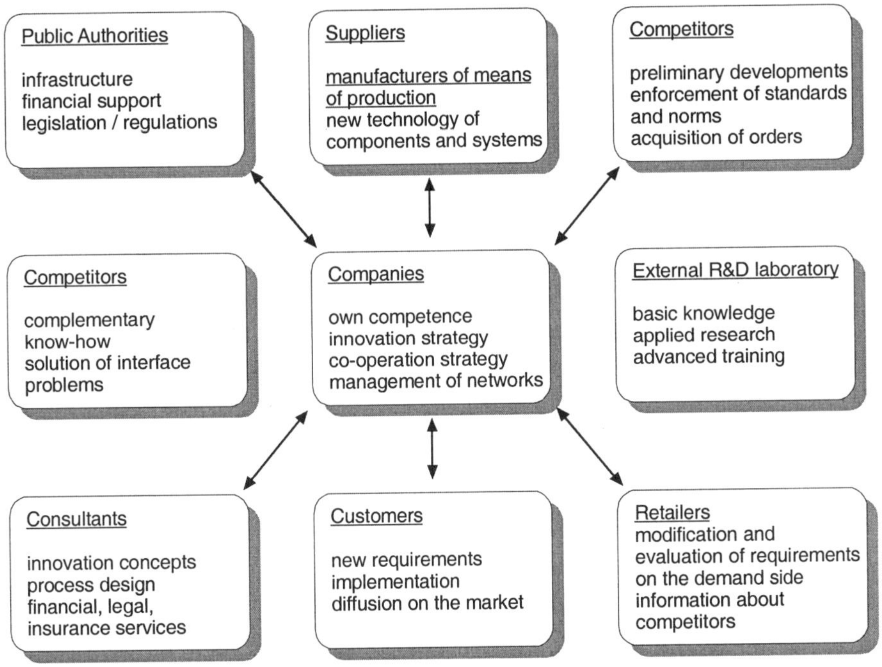

Public Authorities	Suppliers	Competitors
infrastructure financial support legislation / regulations	manufacturers of means of production new technology of components and systems	preliminary developments enforcement of standards and norms acquisition of orders

Competitors	Companies	External R&D laboratory
complementary know-how solution of interface problems	own competence innovation strategy co-operation strategy management of networks	basic knowledge applied research advanced training

Consultants	Customers	Retailers
innovation concepts process design financial, legal, insurance services	new requirements implementation diffusion on the market	modification and evaluation of requirements on the demand side information about competitors

According to Gemünden/Heydebreck (1994: 276)

Networks of different types take shape from these co-operation relationships: they can be identified as "technology clusters" according to their technology fields, as "branch networks" according to their business focus, as "virtual companies" on an entrepreneurial level, and finally as "regional clusters" regarding their spatial location (Hutschenreiter/Penender 1994).

Today, this designation of types seems to be most plausible regarding the actual East German development phase. This is also shown by the first empirical studies (Berteit *et al.* 1998: 12ff., 49-73). Naturally, similar or finer distinctions are also possible, for example according to the degree of inter-dependence in the case of interlocking capital arrangements, formal and informal mechanisms of information exchange, the ability to substitute or complement knowledge acquisition, or according to the conditions of access to or departure from co-operation relationships (cf. Fritsch 1998).

For companies in the new Federal States, the most decisive competitive factors are the accessibility of new technological performance and the utilisation of endogenous potential. The question as to whether, and how this can be managed will also depend on how companies are integrated into networks, with links to suppliers and customers, to investors and to external scientific and research institutions, as well as to R&D service companies.

Due to the fact that relationships are very important both for the access to external resources and for networking activities, the approach to networking models by Gemünden seems to make sense.[1] In accordance with this model, the following four variables are significant for industry:

(1) *Actors:* In industrial innovation networks, companies are dominating actors. Inter-dependent relationships bind these companies to further companies or institutions such as suppliers, competitors for joint system solutions, customers, retailers, consultants, research and training institutes, and public authorities (Gemünden/Heydebreck 1994).

(2) *Resources:* Actors have specific heterogeneous resources at their disposal, which are linked to each other in the net, mostly in a complementary way: human resources, i.e. qualified staff (regarding research and development as well as production, marketing, financing, etc.), and material resources (equipment and production units, financing resources, etc.).

(3) *Activities:* Actors use their resources for specific activities which can include a broad range. In this way, for example, routine and innovation activities can be differentiated. As is regularly shown, network activities are heterogeneous due to different resources (and to different actors). So, networking activities will always mean a certain amount of close co-operation.

(4) *Relationships:* Close co-operation in the net requires close relationships (type of interaction) between the involved actors. These can have varying dimensions with a formal, contractual, legally regulated, or informal character. Interactions concern economic, social, technical, logistical or administrative processes. Although close and long-term relationships are considered as beneficial for the stability of networks, there is a predominance of relationships based on common interests such as profit maximisation or project development.

13.2.2   Definition of Innovation Networks

To describe the state of development of innovation networks in the new Federal States, the following definition seems appropriate: "Here, innovation networks are defined as a system in which different actors (generally legal entities) are linked to each other in order to generate innovations and put them on the market; the actors closely co-operate by bundling mostly complementary resources and by pursuing inter-dependent activities." The development of such networks is not static; instead,

---

[1] Gemünden (1990: 46ff.) (cited according to Gemünden/Heydebreck 1994).

their establishment is dynamic, depending on the level of industrial development, the public R&D infrastructure, innovation potential, as well as on objectives in view of products, sales and profits. Very often, co-operation occurs in response to short-term common interests.

Innovative networks offer the opportunity to bundle different types of capability; thus, advantages related to size are reached, innovation is accelerated, innovation cycles are shortened, and the risk borne by actors in this inter-connected value-added chain, is minimised. This is not only achieved through the inter-linking of goods transport and information transfer. Research and innovation increasingly play an integral role. Although industrial research was very often concentrated on its own objectives and capabilities in the past, an increase is shown regarding the inter-connection of research both between basic and applied research and between individual research processes, via production, up to marketing.

This could lead to

- an improvement of companies' competitiveness due to network synergies such as the advantage of size due to the bundling of capabilities, including facilitated access to markets due to product and process innovation;
- the support of structural change in the individual regions due to the creation of inter-dependent qualified jobs and increased knowledge exchange ("reorganisation through innovation");
- an improved competitiveness of regional applying industries due to the increase of connections with suppliers and the utilisation of existing scientific-technological resources;
- the support of conception development up to concrete planning, as well as the support of targeted settling of companies in areas with modern technology and branches.

Strong impulses for the creation of networks originate through

- the primary availability of new knowledge ("knowledge domain") and its conversion to innovative products and processes, which must take place safe from competitors, and the subsequent market advantage from it;
- regionally motivated privileged access to resources and knowledge assets (industrial districts);
- the sharp competition between system providers and application producers (i.e. which apply their own developments), of very different types, however.

Since the second half of the eighties, interconnected co-operation between small and medium-sized enterprises, as well as external research institutions, has gained in significance. Several factors such as the advance of information and communication technology, the merging of different fields of technology (i.e. mechanics and electronics), the establishment of flexible production systems with a high share of services, and the short-term change in demand behaviour on the side of customers in an increasingly globalised market have led to an acceleration of this development.

13.2.3 Framework Conditions for Innovation Networks in the New Federal States

The crucial framework conditions for innovation networks located in East Germany depend on:

- the state of development and the structure of the East German processing industry, in particular its R&D intensive branches, and the agglomeration of differently sized companies;

- the availability of R&D potential of companies, external industrial research, as well as university and private institutes;

- specialised competence and scientific-technical know-how;

- the quality of the research infrastructure (access to R&D results of university and private research institutes);

- funding opportunities for companies;

- market access problems of innovative companies;

- supra-regional and international co-operation between companies and R&D institutions, as well as

- the effectiveness of promotional measures on both a level of economy and technology policy, aimed at the development of innovative networks in the processing industry according to selected aspects.

Regarding economic renewal in East Germany, industry in particular continues to represent a vulnerable spot. Far-reaching structural changes have occurred during East Germany's transition process towards a free market economy, which has led to completely new conditions regarding research and development as well as economy and especially industry.

For most East German companies, the crucial competitive factor will long remain the gaining of technological knowledge and the closure of the technology gap in order to resolve demand problems and to facilitate market access. Furthermore, given their local orientation, East German companies are still insufficiently integrated into supra-regional networks so that possible spill-over effects do not always reach them.

The achievement of technological competitiveness for East German companies will essentially depend on how they succeed in integrating companies into networks, besides their own research and development activities. Most East German companies have not yet reached a level which allows them to enter or to establish such networks on their own.

Co-operation including internal and external R&D potential, and thus the establishment of innovation networks, is not only seen as a source of new competitive energy but also as a key for organisational modernisation. This, however, is a slow process. It will thus take a long time to establish new efficient networks between companies and research institutions. Assistance herewith, although temporarily and clearly goal-directed, will be one of the future tasks of the States and the Federation. This will be especially important for small companies with few employees and often in fact only for the manager who is able to carry out such scientific-technical tasks.

In addition, more and more complex solutions are required on the markets. Many small and medium-sized enterprises show a lack of appropriate research capacities for this.

According to experience with structural change in modern industries such as the United States, Japan or the former Federal Republic during the 1960s and the 1970s, the branches which benefit from structural change are those which gain competitive advantages over other countries by using production techniques which are human resource intensive. Among these are primarily R&D intensive branches such as mechanical engineering and all branches of electrical engineering.

One of the decisive reasons for the fact that East Germany has not yet accomplished modern structural change is the lack of an industrial environment for East German industrial research.

In East Germany, *industrial research and development* is essentially carried out by small and medium-sized companies. In this field, the lack of large-scale companies doing R&D, which, in the former Federal States, counted among the crucial centres for the establishment of innovation networks, is particularly noticeable. In the new Federal States in 1997, only about 30 % of R&D personnel were employed by companies with more than 500 employees. In 1993, it was still 33.5 %.[2] In the former Federal States in 1997, about 85 % of R&D personnel were employed by companies with more than 500 employees. In 1997, 5 % was accounted for by companies with less than 100 employees in the former Federal States, in the new Federal States 43 % of R&D personnel (in 1993, the number had already been reduced to 38 %).

2　According to data from the Forschungsagentur (Research Agency) Berlin, in 1996 only 23 % R&D personnel were employed by companies with more than 500 employees.

According to a survey carried out by the Forschungsagentur (Research Agency) Berlin, the number of R&D personnel (full time equivalence) in the economic sector of the new Federal States and East Berlin increased again between 1993 and 1998. In 1993, the number was approximately 16,500. In 1998, it had reached about 18,700. The manufacturing industry of the new Federal States and East Berlin showed a decrease in R&D personnel from 27,400 in 1991 to 13,300 in 1993. Up to 1998, this value remained almost constant (approximately 13,870) (Forschungsagentur Berlin 1999).

Scientific statistics according to the Founders' Association (Stifterverband-Wissenschaftsstatistik) show higher figures. This is probably a result of the survey panel used by the Founders' Association's scientific statistics, with a higher share of large-scale companies and the extrapolation based on it (cf. Table 13-1).

Table 13-1: **Total R&D expenditure and R&D personnel of the companies in 1993 and in 1995, according to selected economic branches and to the companies' headquarters**

Economic order	Germany		New Federal States and East Berlin		Former Federal Republic	
	1993	1995	1993	1995	1993	1995
Total R&D Expenditure (in million DM)						
Total	57029	57836	1763	2507	55266	55329
including: Processing industry	53681	55016	1447	2144	52234	52872
including: Chemical industry	10538	10412	217	261	10321	10152
Mechanical engineering	5378	5191	439	546	4939	4645
Automobile construction	12137	13113	29	64	12108	13049
Electrical engineering	13647	13936	247	326	13400	13612
Precision mechanics and optics	829	866	90	131	738	735
Services, as far as delivered by companies	1325	1441	207	304	1117	1136
R&D personnel (full time equivalence)						
Total:	289169	279351	18879	20869	270290	258481
including: Processing industry	276813	264945	15951	17851	260862	247094
including: Chemical industry	53233	48842	2199	2079	51034	46763
Mechanical engineering	34942	31811	4988	4913	29954	26898
Automobile construction	50025	50430	366	612	49659	49818
Electrical engineering	80021	76260	2935	3155	77086	73105
Precision mechanics and optics	5575	5068	1046	1086	4529	3982
Services, as far as delivered by companies	7232	8483	2060	2498	5172	5985

Rounding deviations

Source: Stifterverband Wissenschaftsstatistik/Scientific Statistics by the Founders' Association

A relatively small group of innovative companies with their own R&D personnel (2,800 to 3,000) has been formed in the new Federal States; the members of this group, which shows a tendency for growth, place their stakes on innovations with high risk, partly also at their own R&D expenditure. Their success with the introduction of competitive processes and products is based more and more on networking with other companies, external industrial research institutions, as well as university and private research institutions.

However, the level of expenditure is still very low. Only 2.9 % of the total R&D expenditure of the German economy accounts for East German companies. Compared with the new Federal States, eight times as much financial funding per inhabitant were invested for R&D in the former Federal States (1994: 854 DM vs. 109 DM). West German companies invest 2.3 times as much financial funding per employee and per year for R&D than East German companies do.

In the same way, the East German share of R&D personnel related to 100,000 inhabitants is only one third of the level of former Federal States. Regarding innovative branches such as mechanical engineering, electrical engineering and the chemical industry, East German companies on average only reach about 15 % of the West German level.

Two thirds of the innovative SME, i.e. companies with their own R&D personnel, have permanent relationships with external R&D potential. About 60 % of the companies also maintain permanent or content-related working contacts with external industrial research institutions, universities and colleges. This type of co-operation grows with the size of the company. Whereas about half of the companies with less than 50 employees co-operate with universities and colleges, approximately three quarters of the firms with 200 to 500 employees and larger companies already have 90 % R&D co-operation with universities and colleges. Approximately 30 % of these companies have technical relationships with non-university research institutions (e.g. institutes of the Fraunhofer Gesellschaft, among others). However, the level of these relationships varies. The types of these working relationships range from information exchange and technology transfer by co-workers to joint project realisation. In East Germany, an increasing number of former technical working relationships are being made use of again, which represent a good start for the establishment and further development of innovation networks in the new Federal States.

Due to hive-offs from former collectives and trusteeships, from the Academy of Science, as well as from colleges and new foundations, approximately 240 external industrial research institutions have come into existence; approximately 4,500 people are employed in these hive-offs, from a total of 13,800 R&D personnel employed in the East German processing industry.[3] Two thirds of the R&D personnel of these external industrial research institutes are concentrated in Saxony, Thuringia and East Berlin. Within their economic sector, these external industrial research institutions located in East Germany pursue industrial research in a narrower sense, coupled with development. They also offer further reaching original innovations and solutions to problems, and thus represent the principal endogenous R&D potential for many branches. For this reason they play an important role regarding the establishment of regional and supra-regional innovation networks. Their presence leads to essential competitive advantages and synergies, such as the promotion of start-ups, incentives for the settlement of new companies, and quick technology transfer. Similar to others, East Germany as a country with a high wage level can only survive on supra-regional and international markets with attractive research-intensive products and efficient processes.

External industrial research institutions realise about 40 to 50 % of their order volume through industrial orders, 60 % of which come from the new Federal States, about 35 % from the former Federal States and the remaining 5 % from foreign countries.

In many cases, external industrial research institutions have become crucial centres in the different regions, which promote the settlement of companies with a high technology level, and, thus, of modern industries with qualified employment potential. Some external industrial research institutions not only have outstanding staff potential concerning specific areas of technology, but also development potential for prototypes and partly for small series. They play a particularly important role as initiators of innovation networks. Competence centres for system innovation come into existence around them. Technology centres and technology parks often have a supportive effect.

Although most of the external industrial research institutions have been able to raise their budget shares from economy (some of them already reach 80 to 90 %), at least the preliminary research of these institutes cannot be managed without financial support. Especially the non-profit making external industrial research institutions with a high share of basic research essentially require some kind of fundamental financing. This would not represent additional promotion but would compensate for the lack of legal equation in status with Fraunhofer Institutes, for example, or comparable research institutes from the former Federal States, which are granted basic funding ranging from 20 to 70 %.

Due to the funding granted by the Federal Government and the Federal States Government, the public research field in the new Federal States, which represents a sig-

---

3 SV-Wissenschaftsstatistik (Scientific Statistics by the Founders' Association), survey 1995

nificant basis for the innovative renewal of the economy, has developed well. In round figures, 16 % of the publicly financed R&D personnel in Germany work in the new Federal States. Distinctly more R&D staff work in the university and non-university area than in industry: one R&D employee in colleges or in public research accounts for only 0.7 of an R&D employee in industry. In the former Federal States, the proportion is favourable for industry, since one employee in university and public research accounts for more than two R&D employees in industry. This backlog in East Germany is one of the reasons that East German companies and external industrial research institutions are still less successful than comparable companies and institutions from the former Federal States. There is a strong concentration of university and non-university research institutions in and around Berlin, in Dresden, Jena, Ilmenau, and in a few other regions. Here, the chances for the establishment of innovation networks are particularly good. Due to better connections between industrial research on the one hand, and university and non-university research on the other hand, additional innovation potential could be opened up in East Germany. A high investment amount, as well as the presence of public research institutions, encourage the establishment process of competence centres, which themselves are appropriate starting-points for the establishment and the functioning of networks. The most suitable regions seem to be those of Berlin and its surroundings, Dresden, Jena, Chemnitz and Greifswald. Technology parks can also play a supportive role. This is not yet the case. The first positive results of innovation networks which were established due to technology parks, technology and incubator centres, are to be found in Berlin-Adlershof, Erfurt and Magdeburg.

In the area of complementary research, university and non-university research institutions find themselves increasingly in a position of competition with East German industry. Due to their poor financial situation, especially small and medium-sized East German enterprises often cannot afford to carry out the necessary research or to place orders with research institutions. University and non-university institutions show stronger penetration of the market by reason of the competitive advantages they have, due to their relatively good basic income.

13.3 Statistical Indicators for Innovation Networks

Two criteria from the R&D statistics can be used as indicators for an approach to innovation networks. These are

- the share of externally used R&D expenditure in proportion to the total R&D expenditure of the companies (rate of external R&D expenditure), and

- the share of external order-related R&D financing in proportion to the total R&D expenditure (rate of external R&D financing). Governmental promotion is not taken in account herewith.

On average, a very minimal share of East German companies' total R&D expenditure is utilised as external funds. The absolute amount of external R&D expenditure grew from 130 million DM in 1993 to 171 million DM in 1995. Regarding the former Federal States, the numbers are about 6.4 billion DM for external R&D expenditure in 1993 and approximately 5.6 billion DM in 1995. This indicator distinctly points to the continuing lack of connections with external R&D potential on the side of East German companies. Companies from the former Federal States show an external R&D financing rate of 7.6 % of their total R&D expenditure for 1995, versus 7.8 % in 1993. Companies from the new Federal States show an external R&D financing rate of 4.8 % for 1995 versus 11.5 % in 1993.[4] In 1995, direct government grants for R&D account for 7 % (vs. 6.6 % in 1993) regarding West German companies and for 10.9 % (vs. 13.6 % in 1993) regarding East German companies. Taking into account indirect specific promotions such as, among others, support of R&D staff in East Germany, the government share granted to the appropriate East German companies amounts to about one fifth (24 % in 1993).

- From the external R&D expenditure, i.e. R&D orders placed by East German companies with other companies and with external industrial research institutions, with university and non-university research institutions as well as with foreign research institutions, a little less than 75 % went to the industrial research sector in 1993, 15 % to universities and 7.7 % to non-university institutes. In 1995, proportions changed: About 97 % went to the industrial research sector and only 2 % to public institutes. For the companies from East Berlin and the new Federal States, foreign R&D contractors are almost insignificant with a share of 2.7 % of the external R&D expenditure in 1993, and 3.5 % in 1995. In contrast to this, in the former Federal States more than 18 % of the external R&D expenditure of the industrial sector went to foreign countries in 1993; in 1995 the share was 16 %. In particular West German large-scale companies from the chemical industry, steel construction, mechanical engineering and the automobile industry placed R&D orders abroad. Compared with these figures, East German companies from the corresponding branches only show limited activity abroad. A glance at the chemical industry shows particularly significant differences: In 1993, East German companies only realised 5.1 % of their R&D expenditure with foreign countries; more than 10 times this figure was reached by West German companies with 54.3 %. In 1995, 9.3 % was realised by the East German chemical industry, and 45.2 % by West German companies from the chemical industry.

- The significantly higher external R&D activities on the side of West German companies could also be an indicator of comparatively flat structures. The very distinct experience had in the former Federal States' automobile construction industry shows that the final producer as a "focal company" is interested in the assembly of the least possible pre-manufactured modules. Therefore, a healthy

---

4   Related to companies with external R&D expenditure and/or external R&D financing.

network is established, the co-operation of which is reliable both in terms of supplies and due dates. The suppliers control a net of sub-contractors for accessory parts. Networks of the automobile industry – as in other branches – show that functioning networks are essentially based on common market-directed objectives, on being closely linked to each other, and being controlled and steered on a supra-company level according to their respective production and sales objectives. In the new Federal States, such a level is seldom achieved regarding the establishment of innovation networks. Some large investments, such as BASF-Schwarzheide, Siemens in Dresden and in the Central German "chemistry triangle", represent focal centres for the establishment of networks on the sides of both suppliers and customers; lately, an increasing number of external industrial research institutions, as well as university and non-university research institutes, have also been integrated into these networks. Reference is made to external industrial research institutions such as Chemtec Leuna, located in the chemical triangle of Sachsen-Anhalt, and its close connections to the university institutions of this region. Another example is the steel industry site Eisenhüttenstadt. Beyond the supply and sales connections with their Belgian headquarters, close relationships exist with external industrial research institutions such as IMA Dresden and the Technical University Bergakademie Freiberg.

Another indicator of the establishment of innovation networks are the statements of co-operating companies regarding economic effects.

According to the judgement of entrepreneurs, the shapes of innovating SMEs are clearly reflected both within and outside of networks. In order to have more information about the economic effects of innovation networks eventually coming into existence, some direct questions regarding innovation network approaches were integrated into a written interview of companies. Approximately 1,100 innovating companies were interviewed. It was shown that those SMEs whose capital is inter-linked with other companies (called inter-linked companies) clearly distinguish themselves from non-inter-linked companies in terms of R&D activities, turnover and export rate:

- inter-linked companies have almost four times as many employees as sole traders (on average, the proportion is 153.8 to 40.3 people). In terms of turnover, the inter-linked companies show six times the average of non-inter-linked companies. Consequently, inter-linked companies also show considerably higher turnover productivity.

- With 19.2 % on average, the export share in proportion to turnover amounts to more than twice as much for inter-linked companies as for non-inter-linked companies, which show an export share of 8 %. Related to the sales region "new Federal States", on average 60.7 % of the turnover accounts for sole traders and a little more than one third (36 %) of the turnover accounts for inter-linked companies.

- In 1995, inter-linked companies had on average 13 R&D personnel available (full time or part-time R&D activity), against only 5.4 for sole traders. Regarding full time R&D personnel, an even more favourable proportion is shown for inter-linked companies: with 7.4 employees exclusively pursuing R&D activities, they surpass the level of non-inter-linked firms (2.1 exclusive R&D employees) by more than three times the number. Concerning part time R&D employees, inter-linked companies also show higher average values with 5.7 people versus 3.3 people in non-inter-linked companies.

- Already in 1993, a majority of inter-linked companies (52.3 %) employed full-time R&D staff. This share increased to almost two thirds (62.2 %) by 1996. Sole traders show distinctly less advantageous proportions. In 1993, they only employed a little less than 30 % (29.3 %) full time R&D staff; by 1996, this figure had increased to about one third (37.5 %).

13.4 Perspectives of Economic and Structural Policy

An appropriate means to secure and improve companies' competitiveness is the participation in innovation networks. Obviously, development of innovation networks in the new Federal States is only beginning; at the same time, it is important to realise that the framework conditions for the development of innovation networks in East Germany differ from those in West Germany. A comparison between framework conditions shows the following situation in East Germany:

- There is a lack of large-scale companies acting as motors of the economy, and in particular as nodal points for innovation networks;

- In particular small and medium-sized companies, which represent the industrial majority, have considerable problems in the financing of innovations;

- The promotional means regarding research and innovation are relatively good; they also support the development of innovation networks;

- Although the research structure, i.e. university and non-university research institutions, is well developed, its utilisation by companies is insufficient.

Using the appropriate tools, the government should continue to influence still insufficient framework conditions, which hinder the development of innovation networks in the new Federal States. Basically, giving assistance to this autonomous system should lead to complete independence, i.e. to an ending of governmental funding in the near future.

Some thoughts on this:

In order to enable the creation of competence centres, one of the long-term main objectives of the East German economic system - which is characterised by middle-class companies – should be to inter-link regional innovation potential. These centres could have the potential to become junctions of far reaching innovation networks; at the same time, they could become partners to many small and medium-sized or large-scale companies, as well as to external industrial research institutions. This would also promote the integration of the East German innovation potential into the all-German and international economic system in a sustained way. The more East German contributions are innovative and efficient, the easier their financing and marketing should be, as should be the task of market mechanisms. Consequently, governmental support of the new Federal States should especially focus on the establishment of networks based on top performance technology, which will be required by the German and by the global economy.

The essential task of companies and research institutions is the achievement of economic objectives by innovation-networking with other actors. However, especially the framework conditions designed for East Germany must include the compensation for still existing competitive drawbacks in terms of both technology and location, as well as special encouragement for the opening up of synergy effects.

Therefore, the continuation of both innovation and investment support is considered as necessary in the medium term. The previously existing weak points in the establishment of innovation networks would indirectly be diminished in this process.

In the area of taxation of investment funding, investments within innovation networks should be given additional preference, including industrial research institutions and university and non-university institutes. Since promotional concentration on real capital would be restricted to too short a term, the opening-up of endogenous innovative potential will remain one of the most important tasks in the support of the establishment of a modern industry.

Joint industrial research could play a more important role in view of the establishment of new innovation networks in East Germany. Regarding the use of this promotional measure, access to joint research could eventually be facilitated for small and medium-sized East German companies, and the orientation of university and non-university institutions could be more focused on the tasks of industry.

Encouragement of innovation networks in the new Federal States should particularly focus on the fields of modern technology such as sensor technology, micro-system technology, technology of medicinal equipment, materials technology, and gene technology. Here, special attention is deserved by those innovation networks which come into existence as "virtual companies". This means the "virtual" bundling of competence and capabilities, which then can be realised in a very flexible way, following demand. This leads to considerable cost advantages in all sections of innovation, as well as in marketing logistics.

In special cases, agreements between the federal and regional ministries should enable the concentration of existing promotional funds for the establishment or the further development of regional and supra-regional innovation networks. Concepts by the "Initiative for Machine Tools" Berlin and the "West Saxonian Association of Mechanical Engineering" point in this direction. Therefore, the Federation and the regions should review the funding possibilities of start-up financing; besides investment support, start-up financing should comprise personnel promotion, including qualification and training. This measure would encourage nation-wide acting technology and market oriented competence centres which could produce important innovations for the global market. An important approach was realised by the initiatives "InnoRegio" and "InnoNet", launched by the Federal Ministries of Education and Research and of Economy and Technology.

The East German university and non-university research infrastructure is well developed. However, connections to industrial research and to companies vary strongly according to institutes, companies and regions, and are very weak in many cases. Nevertheless, the level of research co-operation, joint project management, staff exchange and ordered research projects has increased during the past years. The total is still seen as critical. As was confirmed by companies and representatives of university and non-university institutions, many of these organisations are only insufficiently oriented towards industrial tasks. Consequently, teaching and training at university and non-university research institutions should better prepare its graduates to aspire to self-employment in the economy. As an incentive, graduates should be assured of the pre-financing of a hive-off from R&D departments of these institutions, as well as their step into self-employment, during a transitional period.

In many East German technology-intensive companies the function of marketing, which is most important when dealing with innovations, is still insufficiently developed. On a regional level, the possibilities of advanced training should be reviewed. This could be organised by technology and start-up centres.

The establishment of a modern information and communication infrastructure is considered as one of the focal points of developing innovation networks. In this way, the formation of smoothly co-operating teams in innovation networks could be even more efficient.

Complementary to the federal and regional promotional measures for the establishment and development of innovation networks, the various EU framework programmes should be made use of. To inform companies about such promotional options as goal-directedly as possible should be one of the primary objectives. This would save Federal funds and assure a high level of internationalisation to the companies from the new Federal States.

In total, the promotion of industrial research and of the establishment of innovation networks, especially in East Germany, will necessarily be linked to the increase in the number of industrial companies, or the expansion of companies in new expanding branches. This is a long-term process. In East Germany, more commitment on the side of more large investors is needed on the one hand, and innovation and investment cycles of several years in the existing companies, on the other hand. Commitments, however, primarily depend on entrepreneurial evaluation. The settlement of large-scale companies also depends on the available infrastructure. In this area, the existing industrial research and the developing innovation networks play an important role.

13.5 References

BERTEIT, H. ET AL. (1998): Rahmenbedingungen für Innovationsnetze in den neuen Bundesländern und Berlin-Ost, *Materialien zur Wissenschaftsstatistik*, 10. Essen.

BUNDESMINISTERIUM FÜR WIRTSCHAFT UND TECHNOLOGIE (1999): *Wirtschaftsdaten Neue Bundesländer*. Bonn.

DEUTSCHES INSTITUT FÜR WIRTSCHAFTSFORSCHUNG (DIW) (1995): Unternehmerische Netze in der ostdeutschen Industrie: Kooperation hilft überregionale Absatzmöglichkeiten zu erschliessen, *DIW Wochenbericht*, 31.

DIW/IWH/IFW (1999): 19. Anpassungsbericht, *IWH Forschungsreihe* 5/1999. Halle.

DIW/SÖSTRA (1997): *Wirkungen der Programme des BMWi zur Förderung der Industrieforschung auf die Entwicklung des Verarbeitenden Gewerbes in Ostdeutschland*. Berlin.

FORSCHUNGSAGENTUR BERLIN (FAB) (1999): *Längsschnittanalyse der Entwicklung von Innovationspotentialen in Unternehmen der neuen Bundesländer und Vergleich mit der Entwicklung in den alten Bundesländern im Zeitraum 1989-1998*. Neuenhagen b. Berlin.

FRITSCH, M. (1998): Das Innovationssystem Ostdeutschlands: Problemstellung und Überblick, FRITSCH, M./MEYER-KRAHMER, F./PLESCHAK, P. (Eds.) *Innovationen in Ostdeutschland - Potentiale und Probleme*. Heidelberg: Physica-Verlag: pp. 3-19.

GEMÜNDEN, H.G./HEYDEBRECK, P. (1994): *Geschäftsbeziehungen in Netze. Instrumente der Stabilitätssicherung und Innovation*. Wiesbaden.

HUTSCHENREITER, G./PENENDER, M. (1994): Ziele und Methoden der Clusteranalyse wirtschaftlicher und innovativer Aktivitäten, *WIFO Monatsberichte*, 11. Düsseldorf.

MESSNER, D./MEYER, J. (1994): Die Forschungs- und Entwicklungskooperation als Strategische Allianz, *Wirtschaftswissenschaftliche Studien*, 23. Jg., 1/1994.

SEMLINGER, K. ET AL. (1993): Pre-conditions, bottlenecks and problems regarding the creation of new jobs in small and medium-sized companies. Berlin.

WIRTH, S./BAUMANN, A. (1996): Innovative Unternehmens- und Produktionsnetze, TU Chemnitz-Zwickau, *Wissenschaftliche Schriftenreihe*, 8.

WISSENSCHAFTSSTATISTIK GMBH IM STIFTERVERBAND FÜR DIE DEUTSCHE WIRTSCHAFT (1998): Datenreport 1997, Forschung und Entwicklung in der Wirtschaft 1995 bis 1997, *FuE Info* 1/1998. Essen.

14. Innovation Networks and Regional Policy in Europe

Mikel Landabaso*, Christine Oughton, Kevin Morgan

14.1 Introduction: Innovation and Regional Policy

One of the priorities for the new generation of regional development programmes in the European Union for the period 2000-2006 is the promotion of innovation. This is clearly stated in the official Commission Guidelines adopted in June 1999 as the basis for the negotiation of the new generation of regional programmes which should channel to the European regions, less favoured in particular,[1] most of the 213 billion € of the Structural Funds for this period. These Guidelines,[2] entitled "Economic and social cohesion: growth and competitiveness for employment" are based on two broad principles: i) ...identification of integrated strategies for development and conversion and ii) ...the creation of a decentralised, effective and broad partnership. They state that *"Structural assistance should therefore give an increasing priority to promoting RTD and innovation capacities in an integrated manner in all fields of intervention of the Funds"* though actions such as: i) Promoting innovation: new forms of financing (e.g. venture capital) to encourage start-ups, spin-outs/spin-offs, specialised business services, technology transfer, ii) interactions between firms and higher education/research institutes, iii) encourage small firms to carry out RTD for the first-time, iv) networking and industrial co-operation, v) developing human capabilities.

A modified version of this paper was presented at the "3rd International Conference on Technology and Innovation Policy: Global knowledge Partnerships, Creating value for the 21st Century" Austin, USA (August 30 – September 2, 1999).

* The opinions expressed in this paper are the authors' alone and not necessarily those of the European Commission.

1 Over two thirds of this amount is earmarked for regions whose income per capita is less than 75 % of the European average. Nearly 20 % of people in the EU still live in regions with output per head 25 % or more below the EU average. By comparison, just 2 % of the people in the U.S. are in a similar position, and average disparities between States are less than half those between equivalent regions in the EU (CEC 1998b).

2 CEC (1999); CEC (1998a: 12).

The reason for setting this priority within the 2000-2006 Guidelines might lie in the recent evolution of European regional policy due to a new understanding of regional competitiveness (Cooke 1998)[3] and the corresponding matching role of public policy, to which we now turn.

As stated in the Treaty of the Union, European regional policy is mainly about the reduction of disparities among regions in Europe.[4] Thus, European regional policy aims at the creation of the right economic and institutional conditions in a given region for a sustained and a sustainable economic development process which creates economic opportunity and jobs that might increase regional income.

Over and above an appropriate level of physical infrastructures and workforce skills, which have been the traditional target of regional policies, these conditions also involve the existence of regional strengths and opportunities to be further exploited such as the capacity of regional firms to innovate, the quality of management, a business culture which promotes entrepreneurship, an institutional framework which encourages inter-firm and public-private co-operation a dynamic tertiary sector providing business services and the transfer of technology, a minimum level of R&D capabilities, the availability of appropriate interfaces between the demand for and supply of innovation inputs, particularly by/for small firms, and the existence of adequate financial instruments conducive to innovation, etc. These conditions are closely related, at microeconomic level with "intangibles" and "real business services" concepts as opposed to traditional horizontal aid schemes and "automatic" business subsidies.

At the "meso-economic level" they are related to "institutional thickness" and "social capital" concepts. The latter has been defined (Henderson/Morgan 1999) as a relational infrastructure for collective action which requires trust, voice, reciprocity and a disposition to collaborate for mutually beneficial ends. In short, the idea being not to simply to alleviate costs to an individual entrepreneur but to change corporate strategies and business culture as well improving the "productive environment" or "milieu" in which these firms work. This approach can be exemplified in Bellini's

---

3 Cooke (1998: 15-27). This paper summarises the findings of research conducted in the "Regional Innovation Systems: Designing for the future" (REGIS). This research was funded by the European Commission DG 12 under the Fourth Framework – Targeted Socio-Economic Research Programme.

4 The Treaty of the European Union has "economic and social cohesion" as one of its main pillars, as established in the following articles: Article 130a: In order to promote its overall harmonious development, the Community shall develop and pursue its actions leading to the strengthening of its economic and social cohesion. In particular, the Community shall aim at reducing disparities between the levels of development of the various regions and the backwardness of the least favoured regions, including rural areas. Article 130c: The European Regional Development Fund is intended to help to redress the main regional imbalances in the Community through participation in the development and structural adjustment of regions whose development is lagging behind and in the conversion of declining industrial regions.

words: *"the provision of real services transfers to user firms new knowledge and triggers processes within then, thereby modifying in a structural, non transitory way their organisation of production and their relation with the market"* (Bellini 1998).

The legitimacy of public policy for the improvement of these conditions is critically dependent on the assumption that the competitiveness of firms relies not only on its own forces but, in no less extent, on the quality of its environment, sometimes referred to as "structural competitiveness" (Chabbal 1994). The assumption here is that businesses, and SMEs in less favoured regions in particular (mainly because they are working in imperfect markets with limited information and "know-how[5] access") may need assistance in tapping into the necessary resources (related to knowledge, in the form of technology or qualified human capital in particular), to face up to the new forms of competition developing in the global economy. In short they may need more than simply less taxes and lower interest rates to fully exploit their competitive position and thus maximise their contribution to the regional economy in the form of more jobs and higher wealth, which ultimately justifies public financial support for a policy aimed at improving competitiveness. This assumption might particularly hold true in the case of small and medium-sized firms, whose key economic difficulties are related not just to size but also to isolation. And this is ever more true in the case of SMEs working in less favoured regions which are often small size, family owned, working in traditional sectors for local markets and ill prepared for new competitive pressures induced by the globalisation process to which they are increasingly exposed to. Moreover, this assumption is particularly important for regional policy since small and medium-sized firms constitute the basis of the productive fabric of the regions whose development is lagging behind.

Linked to the above, a generally accepted assumption is that the high road to competitiveness for these regions whose firms are progressively exposed to international competition runs through innovation, which enables them to adapt at the right time to increased competitive and the fast pace of technological change. Innovation must apply to all aspects of the activity of a small firm (new markets, new, different or better products, processes and services). In this sense, the concept of innovation embraces research and development, technology, training, marketing and commercial activity, design and quality policy, finance, logistics and the business management required for these various functions to mesh together efficiently.

---

5 We will use von Hippel's definition of know-how as "the accumulated practical skill or expertise that allows one to do something smoothly and efficiently (e.g. the know-how of engineers who develop a firm's products and develop and operate its processes). Firms often consider a large part of this know-how proprietary and protect it as a trade secret".

Since small and medium-sized firms, particularly in less favoured regions, do not usually have either the necessary strategic information or the skills and staff specialising in all the functions listed, some of the latter will have to be carried out by outside contractors. This means that the competitiveness of a small firm depends in part on the quality of the links with and the efficiency/availability of its geographical neighbours (research and technology transfer centres, training centres, business services companies and so on) and it is largely dependent on the quality of the institutional system providing support for innovation (regional authorities responsible for industrial/regional policy in particular). In this sense, innovation is more accessible to small and medium-sized firms when they are working within rich and dynamic regional innovation systems. Regional innovation systems have been defined (Autio 1998) as a distinct concept from national innovation systems (Lundvall 1992), as "essentially social systems, composed of interacting sub-systems; the knowledge application and exploitation subsystem and the knowledge generation and diffusion sub-system. The interactions within and between organisations and sub-systems generate the knowledge flows that drive the evolution of the regional innovation systems". Now, while the core regions of the world economy are well endowed with robust interactive networks, less favoured regions have underdeveloped, fragmented and much less efficient regional innovation systems, as we shall see in the next section.

In short, the creation of the right economic and institutional conditions in a given region for a sustained and a sustainable economic development process implies the triggering of learning processes in the regional economy which allows regional firms to become more innovative, anticipative and adaptable to rapidly evolving markets and techno-economic conditions (cf. Figure 14-1). This is why European regional policy has set innovation promotion as one of the priorities for action in the period 2000-2006, starting from an exploration of new paths by bolstering intangibles,[6] social capital and regional "learning" capacities.

---

[6] This perspective is in line with the "institutional" perspective which insists that these intangible resources merit as much attention as tangible resources (Cooke/Morgan 1998). The so called "institutional perspective" "echoes the bloodless categories of "state" and "market" in favour of a more historically-attuned theoretical approach in which the key issues are the quality of the institutional networks which mediate information exchange and knowledge-creation, the capacity for collective action, the potential for interactive learning and the efficacy of voice mechanisms" (Sabel 1994; Storper 1997; Morgan 1997; Cooke/Morgan 1998; Maskell *et al.* 1998.

Figure 14-1:   The Networked Economy

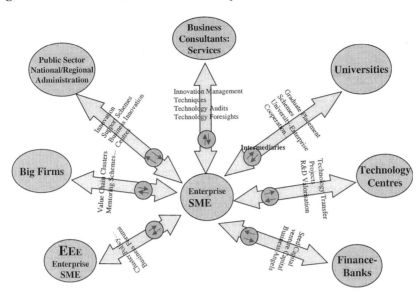

14.2   Regional Innovation Systems and Learning Regions[7]

14.2.1   Regional Innovation Systems in Less Favoured Regions

Today, in Europe, advanced regions spend more public money and in a more strategic way in the promotion of innovation for their firms than less favoured regions. In 1996, for example, while countries such as Denmark, Finland and France spent over 200€ of public aid to R&D per person employed in manufacturing, and Austria, Belgium, Germany and the Netherlands were around the 100€ rate, Greece and Portugal spent 10€ or less, and Spain did not reach 50€ (CEC 1998b). This is increasing the inter-regional innovation gap across Europe, which has a direct relation to the cohesion gap. If regional policy is to be effective in reducing the cohesion gap, it has to address this problem by increasing the innovation capacities in less favoured regions. This, in turn, is dependent on the establishment of an efficient regional innovation system in these regions, as a pre-condition for an increase of public and private investment in the field of innovation.

Otherwise, if policies are solely concerned with increasing the amount of public aid for innovation, "absorption" problems will soon appear and the efficiency of these investments will be undermined, as has already happened in a number of regions

---

7   A more detailed explanation of this section can be found in Landabaso (2000).

with previous policy experiments (e.g. STRIDE). The reason for this lie in what we will call the "regional innovation paradox".

The regional innovation paradox refers to the apparent contradiction between the comparatively greater need to spend on innovation in less favoured regions and their relatively lower capacity to absorb public funds earmarked for the promotion of innovation, compared to more advanced regions. That is, the more innovation is needed in less favoured regions to maintain and increase the competitive position of their firms in a progressively global economy, the more difficult it is to invest effectively and therefore "absorb" public funds for the promotion of innovation in these regions. In other words, one might have expected that once the need is acknowledged/identified (the innovation gap) and the possibility exists, through public means, to respond to it, these regions would have a bigger capability to absorb the resources destined to meet this need, since they start from a very low level ("everything is still to be done"). Instead, these regions face considerable difficulties in absorbing this money. Such is the nature of the regional innovation paradox.

The main cause that explains this apparent paradox is not primarily the availability of public money in the less favoured regions. Its explanation lies elsewhere. It lies in the nature of the regional innovation system and institutional settings to be found in these regions. The regional innovation system in less favoured regions is characterised by its underdevelopment and fragmented nature (cf. Figure 14-2). The institutional setting in less favoured regions is characterised by the absence of the right institutional framework and policy delivery systems, public sector inefficiency and lack of understanding by policy-makers of the regional innovation process in particular. The two combined explain the regional innovation paradox (cf. Table 14-1).

The underdeveloped size of the regional innovation system in less favoured regions and lack of articulation/coherence of its different subsystems and innovation players is illustrated by some of the following characteristics in less favoured regions:

Money earmarked for innovation is sometimes utilised exclusively for the creation of R&D physical infrastructures and equipment for which no real demand has been expressed by the regional firms. Funding might fall in the hands of those responsible for research/science or technology policies which do not have an economic development perspective; innovation being primarily about economic competitiveness and the exploitation of new, better or different markets, products and services. Moreover, the regional government's departments responsible for research and education, industry and economic planning may seldom meet to discuss and agree an integrated policy for the promotion of innovation. That is, there is often no multidisciplinary approach in the planning of funding, which is critically important for a successful innovation policy.

Figure 14-2: **A fragmented regional innovation system: less favoured regions**

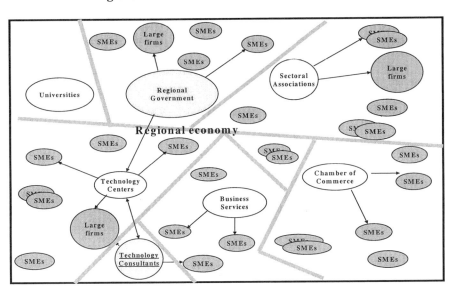

University departments from relatively new universities, for example, which do not have a long tradition of university-industry collaboration, use new funding to strengthen research activities which do not always reflect the needs of the regional firms.

On top of that, regional innovation systems in less favoured regions suffer from isolation from the R&TD networks of "excellence" internationally. Thus, SMEs find it hard to access the technology sources and partners, including informal personal contacts, which are necessary for the continuous feeding of the innovation system in order to keep abreast of technological change in the global economy.

The regional firms, often small, family-owned and competing among themselves in relatively closed markets, do not have a tradition of co-operation and trust either among themselves or with the regional R&TD infrastructure, particularly universities, as illustrated for example by the Spanish case in which "80 % of firms in Spain with fewer than 200 workers undertook no R&D in 1994, whether internally or through outside contractors..." (Fundacion COTEC 1997). Co-operation for innovation which is particularly critical in their case due to their limited internal human resources and "know-how" required for the innovation effort. Firms do not express an innovation demand and the regional R&TD infrastructures are not embedded in the regional economy, and therefore are unable to identify the innovation needs and capabilities existing in the regional economy. Thus, there is a lack of integration between regional supply and demand for innovation.

Table 14-1   Ten structural factors affecting the Regional Innovation Systems in LFRs

1. Shortcomings relating to the capacity of firms in the regions to identify their needs for innovation (and the technical knowledge required to assess them) and lack of structured expression of the latent demand for innovation together with lower quality and quantity of scientific and technological infrastructure.
2. Scarcity or lack of technological intermediaries capable of identifying and "federating" local business demand for innovation (and R&TD) and channelling it towards regional/national/international sources of innovation (and R&DT) which may give response to these demands.
3. Poorly developed financial systems (traditional banking practices) with few funds available for risk or seed capital (and poorly adapted to the terms and risks of the process of innovation in firms) to finance innovation, defined as "long-term intangible industrial investments with an associated high financial risk".
4. Lack of a dynamic business services sector offering services to firms to promote the dissemination of technology in areas where firms have, as a rule, only weak internal resources for the independent development of technological innovation.
5. Weak co-operation links between the public and private sectors, and the lack of an entrepreneurial culture prone to inter-firm co-operation (absence of economies of scale and business critical masses which may make profitable certain local innovation efforts).
6. Sectoral specialisation in traditional industries with little inclination for innovation and predominance of small family firms with weak links to the international market.
7. Small and relatively closed markets with unsophisticated demand, which do not encourage innovation.
8. Little participation in international R&TDI networks, scarcely developed communications networks, difficulties in attracting skilled labour and accessing external knowhow.
9. Few large (multinationals) firms undertaking R&D with poor links with the local economy.
10. Low levels of public assistance for innovation and aid schemes poorly adapted to local SMEs innovation needs

Source: Landabaso 1997

In short, the regional innovation system in these regions does not have either the necessary interfaces and co-operation mechanisms for the supply-demand matching to happen, or the appropriate conditions for the exploitation of synergies and co-operation among the scarce regional R&TD actors which could eventually fill gaps and avoid duplications. In this situation, investing more money in the creation of new technology centres, for example, without previously co-ordinating and adapting the work of existing ones, risks further distorting the system. At the same time it also risks imposing a new budgetary burden on public budgets through the running costs of these institutions, which are unlikely to reach a satisfactory level of self-financing in a reasonable time period due to the mismatch referred before. The same goes for a number of Technology parks initiatives in less favoured regions, which end up becoming property development operations dependent on external capital attraction, poorly linked to the regional industry and playing a very limited role in the economically strategic function of technology transfer regionally.

Moreover, advanced business services and networking agents/interfaces such as those existing in advanced regions are few and not necessarily specialised in the innovation domain. This hinders the innovation opportunities of firms through proper technology auditing and accessing strategically important services such as innovation management, technology forecasting and training, etc. These initiatives, particularly private ones, get trapped in the vicious circle of little demand and poor supply which is rarely spontaneously broken from within the system. When they do respond, due to firms' defensive and adaptive reactions (rather than proactive ones) to market pressures, it is often as late technology followers and innovation opportunities are lost to local industry. Something similar can be said about financial instruments and institutions in less favoured regions, which on top of usually imposing higher than (European) average interest rates offer little attention to long term, higher risk and intangible investments which are characteristic of innovation projects.

Finally, the quality of the institutional setting in these regions is often the main obstacle for the creation of an efficient regional innovation system. Over and above the different degree of regional autonomy in the conduct of regional/industrial policy, several regional governance structures in less favoured regions suffer heavily from lack of credibility, political instability and absence of professional competence (and awareness) in the field of innovation. These three factors are characteristic of underdevelopment.

The lack of credibility of these governance structures, notably vis-à-vis the private sector, is reflected in their limited capacity for consensus building and partnership arrangements with private firms and other institutional actors, be it universities or national R&TD correspondents. Political instability and short-term political consideration (linked to the political cycle) undermine any serious effort in the implementation of an innovation policy which by its own nature is medium to long term. Moreover, it makes the necessary regional leadership for the development of a regional innovation system even more difficult and more prone to fall in the hands of consolidated lobbies and parochial interests which hinder innovation. Lack of professional competence is reflected in the fact that these administrations tend to favour "traditional" and "easy to manage" regional instruments rather than more sophisticated and complex policies such as innovation policy. In some instances even where the political commitment has been clearly expressed to support such a policy, governance structures are often inadequate and it may be difficult to find the necessary management resources to implement it efficiently.

All the above explains to a certain extent the conclusions reached recently by the R&TDI evaluation of Structural Funds for the period 1994-1999 in less favoured regions (Higgings *et al.* 1999: 9) in which the major policy issues identified were:

- Lack of co-ordination between the bodies in charge of public research and those in charge of private research

- Gap between Universities and enterprises

- In many regions there seems to be a lack of co-ordination of the science and technology policy between departments of industry and departments of education

- In some regions there is overlap and inadequate co-ordination between national and regional measures

- There is little involvement of the regional RTDI actors, private sector in particular, in policy planning.

14.2.2   Learning Regions

The innovative capacity of the regional firm is directly related to the "learning" ability of a region. That is, innovative capacity and the regional "learning" ability associated with it is directly related to the density and quality of networking within the regional productive environment. Inter-firm and public-private co-operation and the institutional framework within which these relationships take place are the key sources of regional innovation. Innovation being the end-product and the regional "learning" dependent on the quality and density of the above relationships, being the process.[8]

Asheim (1998: 3) defines a learning region as "representing the territorial and institutional embeddedness of learning organisations and interactive learning" and goes on to argue that in the promotion of such innovation supportive regions the inter-linking of co-operative partnerships ranging from work organisations inside firms to different sectors of society, understood as "regional development coalitions", will be of strategic importance.

A learning region is not a "parochial" region, which ignores the importance of the national and international dimensions, particularly in the fields of "science", "research" and "technology" over and above a narrowly defined concept of "innovation" as such. The regional dimension is important but not exclusive. In this sense it is crucial to acknowledge the need for firms, to be close to "open gates" to the national and international (see Glover 1996) dimensions regionally, in particular for

---

[8] An excellent insight on this "learning" ability can be found in Lundvall/Borras (1997). Chapter 7 in particular, regarding the creation of networks and stimulating interactive learning is most enlightening.

SMEs. Recently, some authors (Koschatzky 1998: 403) have emphasised that even though "space clearly matters in innovation, this takes place more on a perceptive rather than on a politically defined territorial basis" because "it is not a specific region which matters in innovation but an environment fuelled by actors from different regions which, in its complex (inter-regional) structure, has to exceed a critical minimum to be regarded as supportive factor in each region. This environment originates only in part from each single region, but its impact is regionally specific, depending on the structural characteristics of the regional firms". This leads them to conclude that "cross-regional activities would increase the impact of regionally oriented measures, and therefore, provide stronger support for innovation management and the competitiveness of both local and regional firms" (Koschatzky 1998).

Learning as an economic process, can be subject to virtuous circles and increasing returns to scale. The more a region (or a company) is in a position to learn (identify, understand and exploit knowledge, in the form of technological expertise for example, to their own economic benefit) the more capable, and possibly willing, it becomes to build on and increase its demand and capacity to use further new knowledge. But learning depends critically on two key factors; a certain degree of (business-economic) intelligence, which would trigger the demand for new knowledge, and access to/availability of knowledge.[9]

At the meso-economic level we also need an "intelligent cell" to trigger a learning process in a regional economy. The regional government (and its development-related agencies) can play a major role in articulating and dynamising a regional innovation system, understood as the process of generating, diffusing and exploiting knowledge in a given territory with the objective of fostering regional development. In this dynamic and systemic sense, the regional innovation system is in itself the process of learning which "learning regions" are aiming for. The regional innovation system is what determines the effectiveness and the efficiency of regional knowledge building/transfer among the different integrating parts of the system, including individual firms, sectoral/value-chain clusters, business consultants, technology centres, R&D centres, University Departments, laboratories, technology transfer and utilisation of R&D centres, development agencies, etc. The regional innovation system is what makes the whole bigger than the sum of the individual parts (cf. Figure 14-3).

---

9 Similarly, someone that starts reading for the first time, once he/she enjoys a book, looks for more while increasing his-her capacity to understand it better, read faster and combine the new knowledge with previously recorded knowledge from other books, thus extending its learning capabilities in a sort of virtuous circle.

Figure 14-3:   A learning region: An efficient regional innovation system

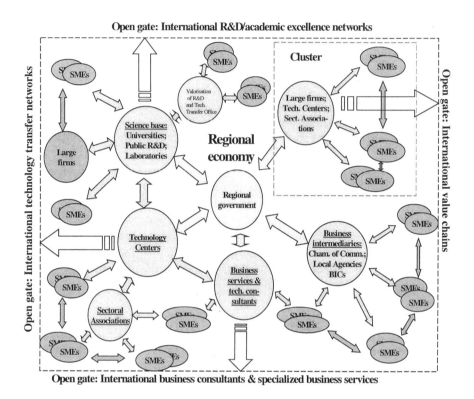

Thus, the regional government can play the role of the "collective intelligence" necessary for a region to spark the process to become a "learning region". It is best placed in terms of political legitimacy and economic powers, including its ability to eventually use the carrot (with for example financial backing: not least as a key decision-maker in the process of Structural Funds allocation) and the stick (for example through its regulatory powers and public procurement policies among others), to facilitate the articulation of the regional innovation system regarding two key aspects in particular. Articulating means linking (regional actors: firms, technology centres, universities, business service providers, etc.) and matching (innovation needs with knowledge supply) in search of synergies and complementarities among the different actors, policies and sub-systems which integrate a regional innovation system. Links, synergies and complementarities which are precisely the "learning vehicles" which may allow a region to effectively learn and increase its innovative potential, due to the nature of the innovation process at regional level.[10]

Firstly by matching innovation (the capacity to use knowledge) demand by firms with existing R&DTI regional supply (the availability of knowledge centres) and eventually finding open gates to external innovation sources and partners capable of addressing the innovation needs of the regional economy. This includes the initial important task of identifying and helping expressing innovation demand and needs, be it latent or not, from regional organisations, most notably SMEs. And secondly, by facilitating co-operation and coherence between the different agents and policies (science policy, research policy, industrial policy, regional policy, human resources policy, competition policy, etc.) which are integral parts of the regional innovation system.

In this sense, the regional government, as evidenced by the RIS experience explained in the next section, can and should play an important role as a catalyst, a facilitator and a broker in the articulation of the regional innovation system. This is particularly important for less favoured regions where the regional innovation system is more fragmented and its subsystems and integral parts are more underdeveloped or, at times, simply completely absent. It is above all a necessary "agent for change" which stimulates and develops networking among the different actors of the regional innovation system in the region. In this "enabling" capacity it can dynamise the regional endogenous potential in terms of entrepreneurship and technical expertise and know-how within the existing business culture and distinctive economic characteristics of the region. Notably by building its own distinctive path to an efficient regional innovation system, since there is not and can not be a unique model of a regional innovation system exportable to all regions. Regional diversity is precisely an asset for regional innovation to build upon.

For the regional government to be able to play the progressive role outlined above regarding the articulation of the regional innovation system, a major cultural and organisational change has to occur in regional governance structures in most regions, and particularly in less favoured regions. This change should go along the lines of more flexible, less bureaucratic structures capable of much tighter partnerships with the private sector (and a higher degree of professional competence in strategic planning capabilities in particular). This also means an increased disposition to consensus building and inclusiveness in the policy process, including the policy delivery system, away from stop and go policy decisions dictated by short-term political instability and parochial interests. It is only then that the necessary "social capital" and "institutional thickness" will be reached in order for the public sector, regional government in particular, to lead the process of articulation and

---

[10] For a further tentative explanation of the characteristics and nature of the innovation process at regional level see Landabaso (1997).

dynamisation of the regional innovation system. That is, the process of learning conducive to the actual realisation of a "learning region" in practice.

Finally, it is important to note that regional "collective learning" takes place in a context of co-opetition (co-operation and competition happening at the same time among the same actors). In this sense, some authors (SRI Consulting 1997: 7) argue that competition in the future may be less between individual firms and more between the value networks (these will include suppliers to the business and other trading partners, even traditional competitors) in which they participate. There will still be competition, but increasingly the participants in the network will also co-ordinate, co-operate, and co-create new opportunities. Trust being at the heart of this horizontal integration process (Sweeney 1999: 19).

This is important from the policy making point of view since it adds a novel role to public action; that of a broker/mediator and facilitator among economic agents in order to create the right conditions for collective learning to happen. In the right context, entrepreneurs could then, through "enlightened self-interest", maximise their contribution to this collective learning task, thus providing further impetus to the broader regional development goals. This has been so far the experience of a number of RIS, as explained in the following section.

14.3   RIS: Towards Collective Learning in Less Favoured Regions

The main objective of innovative actions under the European Regional Development Fund (ERDF) is to influence and improve European regional policy in order to make it more efficient in terms of its content and policy action. These innovative actions rely on "the principle of helping regions to help themselves through initiatives designed to mobilise local knowledge in a process of collective social learning" (Henderson/Morgan 1999).

RISs (Regional Innovation Strategies) are part of these ERDF innovative actions. RISs cost on average half a million € co-financed at 50 % between the EU Commission and the region, and last for two years (for a list of RIS projects cf. Table 14-2). They are not studies or diagnosis of the R&TDI infrastructure of a region in the light of the identified needs of firms. Although they do "use" these studies and diagnosis, they are fundamentally about establishing a socio-economic dynamic (social and institutional engineering) based on a bottom-up open discussion and consensus among the key innovation actors in a region about policy options and new ideas/projects in the field of innovation. In this sense, RIS are also about inter-institutional co-ordination and establishing linkages and collaboration networks among the different elements and players of the regional innovation system. A short definition of RISs might be "an instrument to translate "knowledge" into regional GDP". RIS are a tool to strengthen Regional Innovation Systems (territorial systems that efficiently create, diffuse and exploit knowledge than enhances regional competitiveness) in less favoured regions.

Table 14-2:   Member States and their regions involved in the RIS action

Member State	Region and Objective region eligibility
Austria	Niederösterreich (2&5b)
Belgium	Limburg (2), Wallonie (1&2)
Spain	Aragon (2&5b), Castilla La Mancha (1), Extremadura (1), Galicia (1), Pais Vasco (2&5b), Cantabria (1), Castilla y Leon (1)
Finland & Sweden	Luleå & Oulu (2, 5b, 6)
Greece	Dytiki Makedonia (1), Sterea Ellada (1), Thessaly (1), Epirus (1), Central Makedonia (1)
Germany	Weser-Ems (2&5b), Leipzig-Halle-Dessau (1), Altmark-Harz-Magdeburg (1)
France	Auvergne (2&5b), Lorraine (2)
Ireland	Mid-West, Shannon (1)
Italy	Abruzzo (non-objective), Calabria (1), Puglia (1)
United Kingdom	Strathclyde (2), West Midlands (2&5b), Yorkshire & Humber (2&5b), Wales (2)
Netherlands	Limburg (2)
Portugal	Norte (1), Algarve-Huelva (1)

Within the RIS operation, the Commission also provides regions undertaking a RIS with a network secretariat which facilitates inter-regional co-operation in the form of joint seminars, publications, etc. which promotes cross-fertilisation and the exchange of good practice among participating regions. Furthermore, the Commission also develops a number of "accompanying" measures to enhance the "learning" capacity of participating regions. One of these actions is RINNO. Rinno is intended as a tool for policy-makers to co-operate and learn from each other and avoid reinventing the wheel.

Rinno (which will take the form of a Web Site, CD-Rom and Printed Publication of a Data-Base) has as a key objective the creation and maintenance of an "intelligent" directory of regional public support measures for the promotion of innovation and the identification and diffusion of good practice among regional policy makers. The areas covered by the data base are

(1)   Stimulation and detection of innovation needs in SMEs,

(2)   Support for the development and implementation of innovation projects in SMEs,

(3)   Stimulation and co-ordination of innovation and technology transfer-related business services,

(4)   Linkage mechanisms between the "knowledge base" and regional SMEs.

14.3.1 RIS Objectives: Helping Regions to Help Themselves

RISs have four key objectives:

(1) Place the promotion of innovation as a key priority for the policy agenda of regional governments and develop an innovation culture within regions, particularly, less favoured regions.

(2) Increase the number of innovation projects in firms, particularly SMEs.

(3) Promote public/private and inter-firm co-operation and networks, which facilitate the connection of R&TDI supply with business needs, and the flow of knowledge needed for innovation.

(4) Increase the amount and, more importantly the quality of public spending on innovation through innovation projects, structural funds assistance in particular, and thus promote a more efficient use of scarce public and private resources for the promotion of innovation.

In short, the main objective is to set foundations of an efficient regional innovation system (a "learning" regional economy) by improving existing regional innovation capacities as well as by exploiting the possibilities for new areas of development. RIS focuses on SMEs but is not limited to high-tech sectors and touches upon traditional sectors as well as the service sector (e.g.: tourism) which tend to be important in less favoured regions.

In short, RIS is a "social engineering" action at the regional level whose main aim is to stimulate and manage co-operation links among firms and between firms and the regional R&TDI actors, which may contribute to their competitive position through innovation, notably by facilitating access to "knowledge" sources and partners. In this sense, RIS "social engineering" means creating the right environmental conditions, institutional in particular, for increasing the innovative capacity of the regional economy.

14.3.2 RIS Methodology: a Regional and Demand-Led, Bottom-Up Approach

RISs have six key methodological principles:

• RIS should be based on public-private partnership and consensus (the private sector and the key regional R&TDI players should be closely associated in the development of the strategy and its implementation). Regional administrations should be fully involved, in partnership with the relevant key regional innovation actors, in the design, implementation, monitoring and follow-up of the exercise.

- RIS should be integrated and multidisciplinary: an effort should be made to link efforts and actions from the public sector (EU, national, regional, local) and the private sector towards a common goal. Innovation within RIS includes not only technology considerations but also issues regarding human capital, research and education, training, management, finance, marketing ... as well as policy co-ordination among regional policy, technology policy, industrial policy, R&D and education policy and competition policy.

- RIS should be demand-led (focusing on firm's innovation needs, SMEs in particular) and bottom-up (with a broad involvement of R&TDI regional actors) in their elaboration.

- RIS should be action-oriented and it should include an action plan for implementation with clearly identified projects (at the end of the process new innovation projects in firms and/or new innovation policy schemes and inter-firm networks should appear).

- Regions participating in RIS should exploit the European dimension through inter-regional co-operation and benchmarking of policies and methods.

- RIS should be incremental and cyclical: the exercise is dynamic in the form of a strategy and plan for action that has to be reviewed in the light of previous experience and on-going evaluation.

These principles reflect an approach opposite to one that is top-down, "dirigiste", based on existing institution/power structures and driven by a fiscal transfer/financial distribution rationale, which is characteristic of some of the traditional regional policy stands in a number of less favoured regions. This is in line with the argument that "innovation policy is rather (and increasingly) a matter of networking between heterogeneous (organised) actors instead of top-down decision making and implementation" and it follows that "successful" policymaking normally means compromising through alignment and "re-framing" of stakeholders' perspectives" (Kuhlmann *et al.* 1999: 12).

Moreover, the Commission does not try to promote one standard methodology to be applied religiously in all the regions partaking in RIS projects. In view of the sheer diversity of regional productive environments and their different institutional frameworks, and on the basis of the principle of subsidiarity, the Commission proposes broad guidelines and a flexible methodological approach to regions participating in RIS, which includes:

(1) Raising awareness about innovation and building a regional consensus among key regional actors;

(2) Analysis of the regional innovation system, including technology and market trends assessment, technology foresight and benchmarking with other regions;

(3) Analysis of the strengths and weaknesses of regional firms: assessment of regional demand for innovation services, including technology audits (in SMEs in particular) and surveys regarding firms needs and capacities, including management, finance, technology, training, marketing, etc.;

(4) Assessment of the regional innovation support infrastructures and policy schemes;

(5) Definition of a strategic framework – including a detailed action plan and the establishment of a monitoring and evaluation system. The action plan may involve pilot actions and feasibility studies as well as concrete projects that might be financed under existing structural funds operational programmes.

It is expected that a broad spectrum of local political, economic and academic actors will be involved in this process by actively participating in the Steering Committee responsible for RIS as well as through working groups, seminars, interviews, audits and surveys. In this sense, the suggested institutional setting for carrying forward the RIS is considered to be as important, if not more so than the proposed methodological stages outlined above. That is a Steering Group with broad and active participation of key regional actors and a management unit with the necessary skills (i.e.: economic planning capabilities, business understanding and R&TD competence) together with Working Groups (regional stakeholders which critically review RIS findings and act as a source of innovation projects and new policy approaches) and eventually a process consultant plus regional or international consultants. This institutional setting is essential in order to create the "institutional" dynamism and social engineering that are at the heart of a successful RIS, as we will see in the next section.

Both the principles and the methodology suggested to the regions, which is supposed to be sufficiently compulsory and flexible at the same time so as to provide a clear reference framework while respecting regional diversity of needs and stages of development, are based on a "systemic[11]" vision of the innovation process at regional level.

---

[11] The limitations of the "linear model" of the innovation process have been clearly exposed in the work of Soete/Arundel (1993) by contrasting this model with their "systemic model" of the innovation process. These limitations relate, in particular, to the lack of inter-relation between the different stages and retroactive nature of the innovation process within the "linear model", which is fundamentally a science and technology-push model. It is only by focusing on the demand of firms and the economic nature of the innovation process that policy approaches can deal successfully with the promotion of innovation. Thus, recognising the importance of the economic-pull factors (demand by firms) in the innovation process is critically important in the design of policy.

14.4 The Impact of RIS Projects

In the last five years, more than 600 leading figures in the public and private sectors participated directly in the steering committees of the 32 RIS. The chairmen of most of these steering committees are leading businesspeople (e.g.: Director – Philips International in Limburg (B), Managing Director of Tellabs in Shannon (Ir), Chairman of Surgical Innovations in Yorkshire and the Humber, Managing Director of Wolff Steel Ltd. in Wales, Managing Director NTLCableTel in Strathclyde, etc.) or political figures (e.g.: the presidents of the region in Calabria, and Puglia, Weser-Ems, the regional ministers of industry in Niederösterreich, Castilla La Mancha, Castilla y León, Galicia, Magdeburg, etc., the Secretary General of the Region in Sterea, Thessaly, Central Macedonia, etc.).

More than 5,000 SMEs have been reviewed through technology audits and/or interviews (e.g. 350 firms audited in Wales, a regional innovation survey of 6,000 firms with a 10 % response rate in West Midlands, 1,500 innovation questionnaires sent to regional firms in Thessaly, a survey of 760 companies in Castilla La Mancha with a 18 % response rate and 50 technology audits, a questionnaire survey of 4000 firms with a 15 % response rate in Niederösterreich followed by 30 in-depth firm interviews and 250 companies participating in 5 workshops, etc.).

Several hundred RDTI organisations consulted in the process of drawing up the strategies and in the implementation of action plans based on the RIS (e.g. over 150 businesspeople working in eleven innovation boards in Yorkshire and Humberside, over 200 businesspeople participating in 12 sectoral boards in Castilla y León, 39 experienced innovators from the private sector and 40 academics involved in 17 discussion groups in Shannon, 80 innovation support organisations have worked in RIS Strathclyde, over 150 key regional actors participated in thematic working groups in Calabria, etc.).

What, then, have the RIS achieved so far? One of their most important results has been the increase in and improvement of the human resources required to encourage innovation in the least developed regions and establish networks for organised co-operation.

14.4.1 Expanding Innovation Capabilities through Co-Operation

In many regions the RIS have provided a "shared space for expression" which has helped improve cohesion and institutional co-operation among various socio-economic actors concerned with innovation. For example, the head of the Calabria RIS described the main result of the RIS as being the development of new networks for co-operation among actors based on trust and shared responsibilities,[12] a radically new form of planning in the region, based on consensus and participation, which has also had a marked effect on the new programming of Community regional policy. In that region, as in many other RIS regions, public support for innovation has moved from help for firms and individual projects to start-up aid for "collective topics", i.e. help for firms seeking to co-operate among themselves and encouragement for projects which can develop networks and spread the demand for innovation.

New forms of regional co-operation have developed through RIS, promoting co-operation between firms and between the public and private sectors. In Central Macedonia, for example, the RIS lasted for two years (April 1995-March 1997) and about 200 scientists, public officials and businessmen participated directly in the working groups preparing the 39 reports needed for the Action Plan. Furthermore, 2000 businesses and 277 laboratories participated indirectly during the process of audits, technology demand and supply analyses, and consultation on the selection of projects.

In the Yorkshire and Humberside RIS 11 Sector Innovation Board were established, each with representatives from both large and small firms in the same sector plus representatives from the main business support agencies in the region, TEC and Business Links, universities and further education colleges, local authorities and trade associations. In Aragon six working groups were organised to deal with key regional problems, sectoral matters (logistics, electronics, automation, etc.) and horizontal questions (e.g. financing innovation, innovation in administration). The chairs of these working groups, each of which has about 20 members, are also members of the RIS Steering Committee and half are engaged in business.

So far over 41 funding partners have been involved in the Strathclyde RIS project, in which over 100 organisations from the support network and more than 150 private firms have participated. These networks and new forms of organised co-operation, which have expanded the human resources of the region, have released the energy and creativity required from local actors by identifying new activities and ideas which are needed to increase the number of innovative projects and so help resolve the "regional innovation paradox" mentioned above.

---

[12] On these lines he said that the RIS had been a step towards regarding the public sector not as "an accomplice" but as a "social equal" with public aid becoming a right based on the quality of projects rather than an "income" derived from casual contacts with the political powers.

14.4.2   Greater Transparency in the Regional System for Providing "Knowledge" through Networking

One of the most widespread effects of the RIS has been the establishment of links between the various components of the regional supply of technology: technology centres, polytechnics, technology transfer workshops and so on. The RIS have helped create networks embracing these suppliers of knowledge, mainly in the form of know-how, technology (whether incorporated or not) and technical assistance, so that their respective services gain in complementarity and coherence by building on the possible synergies which exist. In this sense, the creation of a single regional supply network with various access points increases transparency, visibility and understanding in the eyes of the firms which are potential users. It also facilitates public action in this area by avoiding redundancies and identifying gaps, so often boosting the work of regional innovation intermediaries, who are generalists by nature and act as "one-stop shops", so helping strengthen the regional innovation system by improving the links between demand for and supply of regional R&TD, particularly among small firms (cf. Table 14-3).

For example, one of the most tangible results of the RITTS in northern Sweden has been much closer co-operation between the 12 existing "knowledge centres",[13] which are beginning to see themselves as a single multi-centred structure. In Niederösterreich (Austria) encouragement was given for a single network of technological intermediaries, working in complementary fashion with each one concentrating on its basic skills and depending on the support of the others instead of competing against them.

Table 14-3:   Most common RIS generated action in order of priority

1. Creation/strengthening/animation of sectoral business networks, clusters (supply chain or cross-sectoral) and business forums around innovation issues.
2. Establishment of new interfaces between business and the knowledge base, including technology centres, universities, public labs, specialised consultants, etc.
3. Integration and co-ordination of R&TDI services and agencies, including diffusion of their activities vis-à-vis the SME base through "guides", inventories, "one-stop shops", etc. and support to access national/international R&TDI schemes
4. Development of new financial instruments for the financing of innovation (seed corn fund for high tech. start-ups, risk capital, business angels, guarantee funds, etc.) including brokerage services between innovators and the banking sector.
5. Ensuring improvement of market intelligence for forecasting SME technology needs and future leading edge skills needs
6. Identification of innovation projects in firms, SMEs in particular, through the combined efforts of university trainees and R&D labs from Universities and/or other firms
7. Promotion and extension of technology audits in SMEs and innovation management training for businessmen
8. Facilitation of University/big firms spin outs and technology based start-ups

---

13 The knowledge centres generally consist of private firms and local government receiving aid from the ERDF.

The RIS have also had the result of establishing mixed public-private reference groups, both sectoral and otherwise, to investigate the understanding of problems and offer alternative solutions or ways of working. Often the establishment of these networks begins by identifying, establishing and strengthening clusters, skill centres, polis, etc. and in general public-private co-operation networks which bring together the bodies providing knowledge, private firms and the regional or local public sector in order to convert knowledge and "advanced experience" into jobs and new economic opportunities. In this sense we could place these organisations in a general category of "networks" understood as a geographically concentrated social structure within which there is an intensive interchange of knowledge and the development of multi-directional co-operation among members designed to further innovation and economic development.

In this connection it should be noted that, besides the traditional forms of "clusters" or the sectoral groupings encouraged by the RIS, there are new types of "networks", such as the plan for a cross-border "virtual cluster" launched in May 1999 and involving the northern region of Sweden and Finland in the Northern Europe RIS led by the Oulu and Aurorum technology parks.

Examples of measures of this type may be found in the RIS for Yorkshire and Humberside, where 15 "sectoral groupings" were set up to take responsibility for drawing up an innovation plan for each sector. To ensure general co-ordination of the work across the 15 groupings, the Yorkshire and Humberside RIS co-ordinates each "sectoral secretariat" and maintains three horizontal groupings on subjects such as financing innovation, education and business culture and relations between universities and firms. These sectoral groupings have been defined as action-oriented partnerships led by the private sector.

Besides helping to integrate the supply system, these "networks" help reduce or eliminate rivalries between their various constituents since they have a clear demand component led by firms ("business in the driving seat").

Another objective of the establishment of "networks" by some RIS has been to include small firms in the subcontracting chains of large firms. In the Yorkshire and Humberside RIS, for example, the heads of the RIS found that more than 75 % of small firms in the region worked for large firms as subcontractors. The RIS therefore tried to integrate these firms better into the value chain through innovation, so that links with the big "locomotive" firms could provide better value in economic and quality terms. This resulted in the active participation of many senior executives from multinationals based in the region, who even chair some of the "sectoral groupings" there. A similar example may be found in the Plato project in Limburg, where executives from multinationals in the region attend seminars with small businessmen which go beyond specific training to the creation of contacts and the generation of fresh forms of co-operation.

Other interesting examples of this approach include the pilot projects of the Overijsel RIS in the Netherlands. Here the main aim of the measures was to group small firms so that together they could become integrated into the subcontracting chains for large firms. For example, the STUWT project entailed the re-engineering of the subcontracting chain of 80 firms grouped in 13 clusters. Other similar pilot projects are intended to encourage subcontractors to develop into co-producers or principal subcontractors to a large firm by improving quality in order to meet the increasingly strict standards being imposed by the large firms. All this is being achieved by facilitating working in a network with other firms. A grant has been made, for example, to help a large number of small firms to draw up joint proposals to respond to the complex demands of large firms.

14.4.3   Identification of New Innovation Projects in Firms

"In Limburg (Netherlands) 400 companies are taking part in almost 60 projects to date involving the preservation and/or creation of 1,500 to 2,000 jobs. Participation will be intensified from 1999 (aiming at 500 companies per year). Moreover 22 million Ecu from the Structural Funds have been earmarked for RTP projects from the ERDF Objective 2 resources for 1997-1999 and project volume is expected to come to approximately 30 Mecus per annum from 1998 until 2001".

"The Welsh RIS Action Plan, launched by the Secretary of State for Wales in June 1996, includes 66 projects to be led by over 30 different organisations working together in partnership. Of these, over 60 are in progress...".

In the Shannon region the key objective of the RIS is to double the level of innovation by having at least 20 % of all enterprises introducing some new product, process or service in the previous two years by the year 2003. They also want to achieve critical mass with international recognition in at least three networked based sectors within 5 years.

14.4.4   Place the Promotion of Innovation as a Priority for the Regional Policy Agenda and Increase the Amount and, More Importantly the Quality of Innovation Public Spending through Innovation Projects, Structural Funds Assistance in Particular

"In Castilla La Mancha, following the RIS they have increased fivefold the regional budget for innovation promotion from 2.000 million pts for the period 1994-99 up to 15.000 million for the period 2000-2006".

"RIS has clearly become the most important trans-regional co-operation project in central Germany and this impacts extremely positively on its practical implementation, firmly establishing the RIS as one of the main priorities in the regional development programmes, ..." (RTP Leipzig-Halle-Dessau).

"The Welsh RIS has been incorporated into the rationale and project scoring criteria of the Innovation priority of the Industrial South Wales Objective 2 Structural fund programme for 1997-1999, which offers the potential to draw down 18 % of the total programme value of 630 million € to support RTP priorities in South Wales."

"The impact of the RTP in Castilla y León will be perceived on different levels, the first one being - without any doubt - the increase of the region's resources dedicated to the generation of technologies. More specifically, the objective in this sense is that the total R&D expenditure makes up 1 % of the GDP of the region by the year 2001 (nowadays around 0,80 %), similar to the national average estimated by then. In that way, Castilla y León will come closer to the percentages registered in more developed regions and countries. Likewise, another objective to be achieved by the year 2001 is that the R&D expenditure executed by companies represents 50 % of the total R&D expenditure executed in Castilla y León (at present 41,3 %), a number that also lays in the expectations at national level. Finally, by the year 2001, the full-time R&D personnel in companies has to be 40 % of the total number of personnel dedicated to R&D in the region (at the moment 29,8 %), again similar to the national mean foreseen by then".

In 1997 an external evaluation was carried out by Technopolis (Netherlands/UK) and the University of Athens (Greece). The overall conclusion reached by the evaluation team was that "the Regional Technology Plans have had an important impact on the policy formation process, i.e. they created a policy planning culture where innovation and RTD are well embedded in the overall regional development strategies". More specifically the evaluation team pointed to a number of positive results and formulated recommendations for future policy actions in the field of innovation: the strategic nature of the policy approach, which entailed the involvement and co-operation of a broad spectrum of actors in the regional political economy in a detailed planning process, facilitated the development of an endogenous learning environment. It also resulted in a growing awareness of the innovative needs of the regions' firms and therefore instigated a reappraisal of the Structural Fund priorities and spending. Another interesting point to make concerns the flexibility of the RTP and RIS approach. The RTP evaluation demonstrated that despite being applied in many different ways in many different contexts RTPs still had a considerable effect on the regional innovation systems; the model is applicable in dissimilar environments.

In 1999, a second (on-going) evaluation of the RIS took place (Ecotec 1999). The evaluation work was undertaken in 1998 during the RIS funding periods but prior to the 20 RIS pilot projects reaching their final stages. Thus the evaluation does not assess their final impact but rather focuses on aspects of their progress, structures and processes. General conclusion were:

(1) On the Overall Relevance of the Regional Innovation Strategies

"The RIS concept and these objectives are extremely relevant to all of the 20 regions. The underlying problems that the RIS pilot project are designed to address: the relatively poor innovation capacity within the priority region; the weak representation of the business sector and often poorly co-ordinated innovation support structures at the regional level; and, the threat that technological change will adversely impact upon existing economic activity in priority regions, are all evident. This relevance is universal despite the considerable variations in the economic and institutional characteristics between the regions and in the extent and maturity of the existing innovation support activity".

"There has been significant added value at the Community level associated with the RIS pilot projects. Firstly, all projects have embraced the main elements of the RIS concept which themselves stem from experience in different EU member states. Secondly, most projects have made use of and benefited from inputs from consultants and experts from other EU countries than their own. Thirdly, the advice in the published guide, the support of the technical assistance established by the Commission and the networking between the projects have encouraged the adoption of good practice and the transnational exchange of experience ... Finally, the EU funding and the direct link with and obligations to the European Commission at the regional level confers a "leadership role" to the RIS project which can assist in affecting institutional change and co-operation at the local level".

(2) On the effectiveness of the RIS Pilot Projects

"The individual RIS pilot projects appear to be effective in achieving their broad aims. In practice combining the tasks of consensus building, the assessment of demand for and supply of innovation support measures and the preparation of a strategic framework has created synergies and increased overall effectiveness beyond what is likely to have occurred if the strategic frameworks had been developed in isolation..."

"Consensus Building. The achievements with respect to consensus building are evidenced by the level and scope of participation in the pilot projects both formally and informally..."

"Assessing supply and demand for innovation support. The assessments of the regions' assets underpinning innovation have been effective in drawing together and communicating evidence and perspectives on the regions' potential and the constraints upon innovation ...The regions' capacities to undertake future work of this type at the regional, sectoral, thematic and individual firm levels has also increased as a result of the pilot projects".

"Preparing Strategic Frameworks. It can be anticipated with confidence that the RIS pilot projects will generate practical and reasoned strategic frameworks which command a high measure of consensus support. ...Having said this, innovation support to the SME sector is less a matter of overall resource levels and more a matter of the effective use of resources to influence institutional and cultural factors".

"Monitoring and Evaluation. The RIS pilot projects have so far been largely ineffective in developing and implementing monitoring and evaluation systems."

(3) On the lessons emerging

"Lessons for other Structural Fund Innovation Actions: The RIS pilot projects are an example of Article 10 funding being used to "test" a concept in different EU contexts where the concept itself is concerned with identifying how Structural Fund (and other) resources could be better deployed to further an aspect of regional development (i.e. the support of innovation). This approach to the use of the ERDF innovation funds has enabled valuable learning about the validity of the concept as well as the generation of well informed strategic frameworks likely to command consensus support and to improve the deployment of Structural Funds. The approach could be usefully applied to many other "innovative" aspects of regional development and could be extended within the mainstream Structural Funds."

"Lessons for the innovative process and regional development: The RIS pilot projects have helped widen the concept of innovation and strengthen its position within the regional policy agenda. They have also demonstrated how the innovation support system must itself exhibit the characteristics of an innovative and learning organisation. The dual emphasis of the RIS pilot projects on both the structures for involving key actors and the step by step processes for generating strategic frameworks that command consensus support is indicative of this".

"Lessons for consensus and partnership building: The RIS pilot projects have demonstrated how a variety of actors and interested parties can work together in a focussed and progressive manner to appraise and reappraise an important dimension of regional development activity. Each project has needed to take account of its particular institutional, economic and cultural context and to be responsive to the "styles" and outlooks of the individuals involved. However, the consistent focus on the innovation support system has helped generate the conditions for co-operation and commitment from actors in the private sector, the RTD and university communities as well as the public sector, in the process of considering the use of Structural Funds and innovation policy more generally at the regional level".

14.5   Conclusions

A new kind of regional policy is emerging in the European Union in which the accent is on collective learning and institutional innovation rather than upon basic infrastructure provision. Thus emphasis is being put on "social capital" (i.e. a relational infrastructure for collective action based on trust, reciprocity and the disposition to collaborate to achieve mutually beneficial ends) rather than into "physical capital" building.

In our opinion, in these small regional experiments we can begin to discern a new and more innovative form of economic governance, the hallmarks of which are interactive, strongly based on public-private partnership and network based, rather than hierarchical of solely market-driven, the respective governance modes of dirigisme and neo-liberalism. In this sense these actions are concerned not with the scale of state intervention but its mode, not the boundary between state and market but the framework of effective interaction (Henderson/Morgan 1999). In RIS governance is based on public-private partnership and consensus, with the public sector playing a role mainly as an animator, a catalyst and a dynamic force for networking among all the relevant regional agents. Thus making scarce regional efforts and energies converge towards innovation promotion with a regional development objective.

Although regional experiments like RIS have triggered some encouraging institutional innovations, the key question is how to evaluate success in this context. It is clear that this new regional policies, which aim to raise regional innovation capacity, and the establishment of an efficient regional innovation system, can not be judged by the standards of the old regional policies (short-term job creation). In order to overcome this problem, we need a new set of indicators that allow us to assess longer-term changes in regional innovation capacity, away from "linear" indicators such as standard R&D input and output indicators. That is, we need interactive indicators which aim to measure "soft" processes like institutional linkages and network formation in order to capture the important changes in a region's institutional architecture which are beyond the grasp of more conventional linear indicators (Nauwelaers/Reid 1995).

Notwithstanding this lack of appropriate indicators, the external evaluation of the first 7 RIS experiences, as well as the on-going evaluation carried out for the new generation of 20 RIS, clearly point out that RIS have been a novel approach in each of the regions involved. An approach that has significantly contributed to establishing a strategic planning culture based on consensus and partnership by which innovation promotion has been put high in the regional development agenda. Moreover, most regions which have undertaken a RIS have managed to increase considerably the quantity of public funds targeted on innovation, which is expected to trigger a parallel increase from private funds.

In conclusion, we believe that RISs may help prepare the ground so that those responsible for innovation promotion at regional level can better respond to the need of increasing the regional innovation potential and addressing the problem of "absorption" related to the "regional innovation paradox" mentioned before. This can be done specifically through work on basic strategic planning involving key regional actors which will result in new innovation projects consistent with regional policy objectives. Social capital, learning (inter-firm, inter-organisational and inter-regional) and networking to promote collective external economies of scale are crucial to this process. Thus, RIS seems to be a fertile ground for further experimentation and learning in the quest for efficient regional innovation systems that can consolidate sustained and sustainable economic development processes in those regions where these are most needed.

14.6   References

ASHEIM, B.T. (1998): *Learning regions as development coalitions: partnership as governance in European workfare states?* Paper presented at the second European Urban and Regional Studies Conference on "Culture, place and space in contemporary Europe", University of Durham, UK, 17–20th September 1998.

AUTIO, E. (1998): Evaluation of R&TD in Regional Systems of Innovation, *European Planning Studies*, 6, pp. 131-140.

BELLINI, N. (1998): *Services to Industry in the Framework of Regional and local Industrial Policy.* Draft paper OECD Modena Conference (May 28-29 1998) on "Up-grading knowledge and diffusing technology to small firms: building competitive regional environments".

CEC (1998a): *Seventh Survey on State Aid in the manufacturing and certain other sectors.* SEC (99) 148. Brussels: EU Commission.

CEC (1998b): *Sixth Periodic Report on the Social and Economic Situation and Development of the Regions of the European Union.* Brussels: EU Commission.

CEC (1999): *The Structural Funds and their Co-ordination with the cohesion fund: guidance for programmes in the period 2000-2006 tabled by Mrs. Wulf-Mathies in agreement with Mr. Flynn, Mr. Fischler and Ms Bonino*, Brussels. (www.inforegio.cec.eu.int/wbdoc/docoffic/coordfon/coord_en.htm).

CHABBAL, R. (1994): *OECD Programme on Technology and Economy*. OECD Paper. Paris: OECD.

COOKE, P. (1998): The role of Innovation in Regional Competitiveness, COBBENHAGEN, J. (Ed.) *Cohesion, Competitiveness and RTDI: their impact on regions*. Conference Proceedings, Maastricht, The Netherlands, May 28 and 29, 1998.

COOKE, P./MORGAN, K. (1998): *The Associational Economy: Firms, Regions and Innovation*. Oxford: Oxford University Press.

ECOTEC (1999): *The On-going evaluation of Regional Innovation Strategies Pilot Projects*. Brussels: EU Commission.

FUNDACION COTEC (1997): *Tecnología e Innovación en España; Report 1997*. Madrid.

GLOVER, R.W. (1996): The German apprenticeship system: lessons for Austin, Texas, *Annals of the American Academy of Political and Social Science*, 544, pp. 83-94.

HENDERSON, D./MORGAN, K. (1999): Regions as laboratories: the rise of regional experimentalism in Europe, WOLFE, D./GERTLER, M. (Eds.) *Innovation and Social Learning*. Macmillan, St. Martins Press, North America.

HIGGINGS, T. ET AL. (1999): *R&TDI Thematic Evaluation of the Structural Funds 1994-1999 – Spain country report*.

KOSCHATZKY, K. (1998): Firm innovation and region: the role of space in innovation processes, *International Journal of Innovation Management*, 2, pp. 383-408.

KUHLMAN, S. ET AL. (1999): *Improving Distributed Intelligence in Complex Innovation Systems*, Final report of the Advanced Science & Technology Policy Planning Network, Karlsruhe, Fraunhofer ISI.

LANBABASO, M. (1997): The promotion of Innovation in regional policy: proposals for a regional innovation strategy, *Entrepreneurship & Regional Development*, 9, pp. 1-24.

LANDABASO, M. (2000): EU Policy on innovation and regional development, BOEKEMA, F./MORGAN, K./BAKKERS, S./RUTTEN, R. (Eds.) *Knowledge, innovation and economic growth: the theory and practice of learning regions*. Cheltenham: Edward Elgar, pp. 73-94.

LUNDVALL, B.-Å. (Ed.) (1992): *National Systems of Innovation: towards a theory of innovation and interactive learning*. London: Pinter Publishers.

LUNDVALL, B.-Å./BORRAS, S. (1997): *The globalising learning economy: implications for technology policy*. Final Report under the TSER Programme. Brussels: EU Commission.

MASKELL, P./ESKELINEN, H./HANNIBALSSON, I./MALMBERG, A./VARNE, E. (1998): *Competitiveness, Localised Learning and Regional Development. Specialisation and Prosperity in Small Open Economies*. London: Routledge (=Routledge frontiers of political economy 14).

MORGAN, K. (1997): The learning region: institutions, innovation and regional renewal, *Regional Studies*, 31, pp. 491-503.

NAUWELAERS, C./REID, A. (1995): *Innovative Regions*. Brussels: European Commission.

RIS Aragón Final Report RIS+ proposal, EU Commission 1999.

RIS Calabria Final Report RIS+ proposal, EU Commission 1999.

RIS Castilla La Mancha Final Report RIS+ proposal, EU Commission 1999.

RIS Dytiki Makedonia Final Report RIS+ proposal, EU Commission 1999.

RIS Galicia Final Report RIS+ proposal, EU Commission 1999.

RIS Niederosterreich Final Report RIS+ proposal, EU Commission 1999.

RIS Norte Final Report, EU Commission 1999.

RIS North Sweden Final Report RIS+ proposal, EU Commission 1999.

RIS Pais Vasco Final Report RIS+ proposal, EU Commission 1999.

RIS Shannon, Final Report. EU Commission 1999.

RIS Strathclyde Final Report RIS+ proposal, EU Commission 1999.

RIS Thessaly, Final Report RIS+ proposal, EU Commission 1999.

RIS Weser-Ems Final Report RIS+ proposal, EU Commission 1999.

RIS West Midlands Final Report RIS+ proposal, EU Commission 1999.

RIS Yorkshire and Humberside Final Report RIS+ proposal, EU Commission 1999.

RITTS Canarias Final Report RIS+ proposal, EU Commission 1999.

RITTS Overijssel Final Report RIS+ proposal, EU Commission 1999.

RITTS Umbria Final Report RIS+ proposal, EU Commission 1999.

RTP Castilla y Leon Final Report RIS+ proposal, EU Commission 1999.

RTP Central Makedonia Final Report RIS+ proposal, EU Commission 1999.

RTP Leipzig-Halle-Dessau Final Report RIS+ proposal, EU Commission 1999.

RTP Limburg Final Report RIS+ proposal, EU Commission 1999.

RTP Lorraine Final Report, EU Commission 1997.

RTP Northern EU Final Report RIS+ proposal, EU Commission 1999.

SABEL, C.F. (1994): Flexible Specialisation and Re-emergence of Regional Economies, AMIN, A. (Ed.) *Post-Fordism. A Reader*. Oxford: Blackwell, pp. 101-156.

SOETE, L./ARUNDEL, A. (Eds.) (1993): *An integrated approach to European Innovation and Technology Diffusion Policy: A Maastricht Memorandum*. Brussels: CEC.

SRI CONSULTING (1997): *A new formula for competitiveness: trust*. BIT3M Futurescript. London: SRI Consulting Ltd.

STORPER, M. (1997): *The Regional World. Territorial Development in a Global Economy*. New York: Guilford Press.

SWEENEY, G. (1999): *Local and regional innovation: governance issues in technological, economic and social change*. Report of the Six Countries Programme conference in Ireland 1997.

VON HIPPEL, E. (1998) The Sources of Innovation. New York: Oxford University Press.

List of Contributors

Herbert Berteit   SÖSTRA Forschungs-GmbH
                         Berlin
                         Germany
                         berteit@soestra.de

Ulrike Bross   Booz, Allen & Hamilton
                         Düsseldorf
                         Germany
                         bross_ulrike@bah.com

Antoine Bureth   Bureau d'Economie Théorique et Appliquée (BETA)
                         Université Louis Pasteur
                         Strasbourg
                         France
                         bureth@cournot.u-strasbg.fr

Michael Fritsch   Technical University Bergakademie Freiberg
                         Faculty of Economics and Business Administration
                         Freiberg
                         Germany
                         fritschm@vwl.tu-freiberg.de

Jean-Alain Héraud   Bureau d'Economie Théorique et Appliquée (BETA)
                         Université Louis Pasteur
                         Strasbourg
                         France
                         heraud@cournot.u-strasbg.fr

Knut Koschatzky   Fraunhofer Institute for Systems and Innovation Research
                         Karlsruhe
                         Germany
                         ko@isi.fhg.de

Marianne Kulicke   Fraunhofer Institute for Systems and Innovation Research
                         Karlsruhe
                         Germany
                         mt@isi.fhg.de

Mikel Landabaso   EC Commission DG REGIO (CSM2 1/158)
                         Bruxelles
                         Belgium
                         Mikel.Landabaso@cec.eu.int

Kevin Morgan   Dpt. of City and Regional Planning
                         Cardiff University
                         Cardiff, Wales
                         Great Britain

Emmanuel Muller

Fraunhofer Institute for Systems and Innovation Research
Karlsruhe
Germany
em@isi.fhg.de

Christine Oughton

Birkbeck College
University of London
London
Great Britain
c.oughton@bbk.ac.uk

Franz Pleschak

ISI-Research Department Innovation Economics
Technical University Bergakademie Freiberg
Freiberg
Germany
pleschak@bwl.tu-freiberg.de

Slavo Radosevic

University College London
School of Slavonic and East European Studies
London
Great Britain
s.radosevic@ssees.ac.uk

Javier Revilla Diez

University of Hanover
Department for Economic Geography
Hannover
Germany
diez@mbox.wigeo.uni-hannover.de

Simone Strambach

University of Stuttgart
Geographical Institute
Stuttgart
Germany
simone.strambach@po.uni-stuttgart.de

Frank Stummer

ISI-Research Department Innovation Economics
Technical University Bergakademie Freiberg
Freiberg
Germany
stummer@merkur.hrz.tu-freiberg.de

Günter H. Walter

Fraunhofer Institute for Systems and Innovation Research
Karlsruhe
Germany
wa@isi.fhg.de

Andrea Zenker

Fraunhofer Institute for Systems and Innovation Research
Karlsruhe
Germany
az@isi.fhg.de

Index

G

H

I

J

K

L

M

N

O

P

Q

R

Druck: Strauss Offsetdruck, Mörlenbach
Verarbeitung: Schäffer, Grünstadt